W9-DGU-589

Macmillan
Revised

ENCYCLOPEDIA
of SCIENCE

The 3 Earth

Our Planet –
Its Land, Sea, and Air

Dougal Dixon

Macmillan Reference USA
New York

Board of Advisors
Linda K. Baloun, B.S., Science Teacher, Highmore, SD

Merriley Borell, Ph.D., Consultant Contemporary Issues in Science and Medicine, Piedmont, CA

Maura C. Flannery, Ph.D., Associate Professor of Biology, St. John's University, Jamaica, NY

Owen Gingerich, Ph.D., Professor of Astronomy and History of Science, Harvard-Smithsonian Center for Astrophysics, Cambridge, MA

MaryLouise Kleman, M.A., Science Teacher, Baltimore County, MD

Irwin Tobias, Ph.D., Professor of Chemistry, Rutgers University, New Brunswick, NJ

Bibliographer
Sondra F. Nachbar, M.L.S., Librarian-in-Charge, The Bronx High School of Science, Bronx, NY

Advisory Series Editor
Robin Kerrod

Consultants
Dr. J. Beynon
Jeremy Bloomfield

Andromeda Staff
Caroline Sheldrick, Editor
Tamsin Osler, Assistant Editor
David West/Children's Book Design and
Ian Winton, Designers

Front Cover: Industrial zone (*Hutchison Library/ Bernard Regent*)

Published by:
Macmillan Library Reference
A Division of Macmillan Publishing USA
1633 Broadway, New York, NY 10019

Copyright © 1991, 1997 Andromeda Oxford Limited
An Andromeda Book
Devised and produced by Andromeda Oxford Ltd,
11-15 The Vineyard, Abingdon, Oxfordshire, OX14 3PX England

Macmillan edition copyright © 1991, 1997
Macmillan Reference USA

Library of Congress Cataloging-in-Publication Data

Macmillan encyclopedia of science. -- Rev. ed.
     p.   cm.
    Includes bibliographical references and index.
    Summary: An encyclopedia of science and technology, covering such areas as the Earth, astronomy, plants and animals, medicine, the environment, manufacturing, communication, and transportation.
  ISBN 0-02-864556-1 (set)
  1. Science--Encyclopedias, Juvenile. 2. Engineering--Encyclopedias, Juvenile. 3. Technology--Encyclopedias, Juvenile. [1. Science--Encyclopedias. 2. Technology--Encyclopedias.]
I. Title: Encyclopedia of science.
0121 M27 1997
500--DC20   96-36597
                               CIP
                               AC

Volumes of the *Macmillan Encyclopedia of Science*
Set ISBN 0-02-8645561
  1 *Matter and Energy* ISBN 002864557X
  2 *The Heavens* ISBN 0028645588
  3 *The Earth* ISBN 0028645596
  4 *Life on Earth* ISBN 002864560X
  5 *Plants and Animals* ISBN 0028645618
  6 *Body and Health* ISBN 0028645626
  7 *The Environment* ISBN 0028645634
  8 *Industry* ISBN 0028645642
  9 *Fuel and Power* ISBN 0028645650
10 *Transportation* ISBN 0028645669
11 *Communication* ISBN 0028645677
12 *Tools and Tomorrow* ISBN 0028645685

Printed in the United States of America

Introduction

This volume deals with geology and with two aspects of the Earth's surface that enable it to support life: its abundant water resources and the atmosphere that covers it.

To learn about a specific topic, start by consulting the Index at the end of the book. You can find all the references in the encyclopedia to the topic by turning to the final Index, covering all 12 volumes, located in Volume 12.

If you come across an unfamiliar word while using this book, the Glossary may be of help. A list of key abbreviations can be found on page 87, and a diagram of eras and periods in the Earth's history on page 29. If you want to learn more about the subjects covered in the book, the Further Reading section is a good place to begin.

Scientists tend to express measurements in units belonging to the "International System," which incorporates metric units. This encyclopedia accordingly uses metric units (with American equivalents also given in the main text). In illustrations, to save space, numbers are sometimes presented in "exponential" form, such as 10^9. More information on numbers is provided on page 87.

Contents

Part One

Our changing Earth

The Earth was born from a cloud of dust and gas nearly 5 billion years ago, along with the other planets in our Solar System. But the Earth evolved quite differently from the others, making it possible for life to flourish in a myriad of different forms and environments.

From the time that the Earth cooled and formed a solid crust, it has been ever changing. Even as the rocks and mountains formed, the weather and other agents set to work to grind them down. Huge continents formed and then began to split up under the action of relentless forces deep inside the crust.

Such elemental forces are still at work today, creating and demolishing mountains, and setting the Earth quaking and volcanoes erupting. The investigation of such forces forms part of one of the most exciting of all the sciences, geology.

◄ The Grand Canyon in Arizona. The canyon was formed over millions of years. It has a maximum depth of some 2,000 m (well over 1 mi.) from the plateau top to the Colorado River.

The planet Earth

The Earth, our home planet, is quite unlike any other in the Solar System. It was formed at the same time as the rest of the Solar System. It is a rocky planet like Mercury, Venus, and Mars, the other inner planets. However, the Earth's distance from the Sun gives it the conditions that can allow water to exist and first enabled life to develop. The Earth spins on its axis and moves around the Sun. These motions give us, respectively, days and years. The tilt of its axis creates the seasons. Modern scientific techniques allow us both to investigate the Earth's interior and to survey the surface of the planet.

SPOT FACTS

• The volume of the Earth is about 1 sextillion cu m ("1" followed by 21 zeroes); this is about 1.3 sextillion cu. yd.

• The density of the Earth is about 5.5 times the density of water.

• The Earth's mass is about 6 septillion kg ("6" followed by 24 zeroes); this is about 13 septillion lb.

• The average distance of the Earth from the Sun is 149,503,000 km (92,897,000 mi.).

• The Earth is not a perfect sphere but is slightly pear-shaped: the Equator bulges 21 km (13 mi.); the North Pole bulges 10 m (33 ft.); and the South Pole is depressed about 31 m (about 100 ft.).

• The total surface area of the planet is about 510 million sq km (197 million sq. mi.).

• The highest recorded surface temperature on Earth was 58°C (136°F) in Libya. The lowest was -90°C (-129°F), recorded in Antarctica. The average surface temperature is 14°C (57°F).

• Along with the rest of the Solar System, the Earth is traveling through space toward the constellation of Hercules at about 20.1 km/s or 72,360 km/h (about 12.5 mps or 45,000 mph).

• The whole Milky Way galaxy is moving toward the constellation of Leo at approximately 600 km/s (about 375 mps).

• The Earth, together with the Moon, moves in an orbit about the Sun. The orbit is almost circular. The length of the Earth's orbit is about 938,900,000 km (583,400,000 mi.) and the Earth travels along it at a speed of approximately 106,000 km/h (about 66,000 mph).

• Because the distance around the Earth is greater at the middle than at the ends, different parts of the planet appear to rotate at different speeds. A point on the Equator moves at a speed of about 1,600 km/h (about 1,000 mph), and a point on the Earth at the latitude of Halifax, Nova Scotia (45° North), moves at a speed of about 1,073 km/h (about 667 mph).

• The Earth consists of different concentric shells, like the layers of an onion. The main layers are the crust, the mantle, and the core.

• Only 1/400th of the Earth's mass consists of continental crust, which we are able to examine. Continental crust is on average at least four times thicker than oceanic crust, which is extremely difficult to examine.

• Convection currents within the Earth's mantle are thought to be the driving force behind continental drift and seafloor spreading.

• Convection currents in the iron/nickel core may act as a dynamo, generating the Earth's magnetic field.

• The radius of the Earth's core is about 3,500 km (2,175 mi.). The core may reach temperatures as high as 6,650°C (12,000°F).

The unique Earth

The Sun is at the center of a whole family of planets. Relatively close to it are the inner planets – Mercury, Venus, Earth, and Mars.

Mercury, closest to the Sun, is a cratered rock not much larger than our Moon. It is a blisteringly hot planet totally incapable of supporting life of any kind. Mercury has an extremely thin atmosphere that contains largely sodium and potassium.

Venus is the next planet out from the Sun and is about the same size as the Earth. In contrast to Mercury, Venus is clothed in a thick atmosphere of mainly carbon dioxide. At the surface, the pressure of the atmosphere is nearly 100 times that on Earth. The thick blanket of gases acts like a greenhouse and traps the Sun's heat, and gives a surface temperature as high as $480°C$ ($900°F$). There can be no life of any kind on the planet Venus.

Mars, the first planet beyond us, is smaller than the Earth. Mars also has an atmosphere, but only a very-low-pressure one. Again, the atmosphere consists largely of carbon dioxide.

Water exists on the surface, but only in the form of ice at the poles. Pressures and temperatures are never high enough to make all the ice melt. The whole planet is a red, lifeless desert.

Beyond Mars lie the outer planets – the gas giants of Jupiter, Saturn, Uranus, and Neptune. These are far larger than the Earth, and their outer visible structure consists of the gases hydrogen and helium. They are so unlike the Earth and the other inner planets, in structure and composition, that it is difficult to make comparisons. Beyond them, the outermost planet Pluto is so distant that it is something of a mystery altogether.

Among all the planets, only the Earth is of such a size and at such a distance from the Sun that its surface is neither very cold nor very hot. Because of the temperature and pressure, water can exist in all its three forms – as gas, liquid, and solid. These conditions also allow plants and animals to reproduce and evolve. In short, Earth is the only planet that has the conditions that support life.

▲ Photographs sent back from the Martian surface by the *Viking* space probes in 1976 showed a red, stony desert with a red, dusty sky. No sign of life was found. The low temperatures and pressures make Mars very unsuitable for living things.

◀ The Earth has blue skies filled with clouds, and in most places there is standing or running water. Life abounds, with plants using sunlight and water to make food. Plants in turn support the complex web of animal life across the globe.

9

Earth's origins

Since the dawn of history people have wondered about how the Earth came to exist. Many theories were put forward, based on what little knowledge was available at the time. Most of these are now out of date. Nowadays, with the use of modern technology, we are continually amassing more and more information about the Solar System and how the Earth was born. We use satellites, space probes, and scientific instruments to look at the stars and the planets. We are gradually finding out what makes the stars give out light and what the planets are made of. Scientists use this knowledge to find out more about our own Earth.

The most up-to-date theory suggests that the whole Solar System – the Sun, the Earth and the rest of the planets, moons, and asteroids –

formed from a single mass of dust and gas called a nebula. About 4.7 billion years ago the nebula began to shrink. This was caused by the force of gravity, which pulls objects together because of their mass.

While this process continued, the Sun began to form. As it shrank, the nebula began to spin, which made it flatten out into a disk. The spinning at the center was greater than toward the outside of the disk, and the outermost parts sheared off as rings. The material at the center heated up rapidly, and then began to emit energy as the Sun.

The planets could have formed when the gas and dust in the rings gathered together into solid bodies. Or they may have been built up from layers of dust.

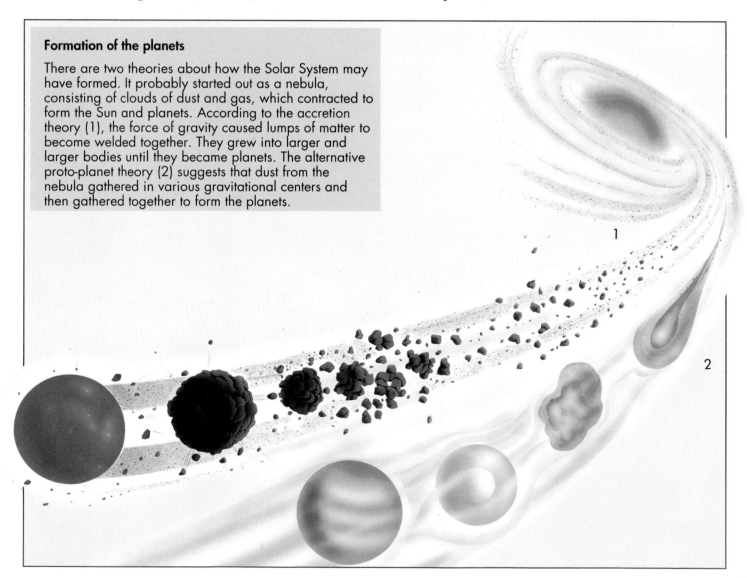

Formation of the planets

There are two theories about how the Solar System may have formed. It probably started out as a nebula, consisting of clouds of dust and gas, which contracted to form the Sun and planets. According to the accretion theory (1), the force of gravity caused lumps of matter to become welded together. They grew into larger and larger bodies until they became planets. The alternative proto-planet theory (2) suggests that dust from the nebula gathered in various gravitational centers and then gathered together to form the planets.

10

Gravity and magnetism

A tremendous force caused the original nebula to contract and form the Solar System. This was the force of gravity. It is difficult to define, but it acts on all matter in the Universe. Its effect is to pull toward each other all things that have mass – planets, rocks, or particles of dust. The gravitational force between the Earth and all of the objects on the Earth causes the objects to have weight.

Other major effects result from the Earth's magnetic field. The Earth acts like a giant magnet, tending to pull magnetic materials toward its north and south poles. This effect is used in a compass, which points northward because the magnetic north pole is near the geographical North Pole. The origin of the magnetic field may lie in the Earth's core.

▲ The Aurora Borealis, or Northern Lights, occurs when the Earth's magnetic field traps charged particles from the Sun. The particles interact with molecules in the air, and this makes them give off a glow.

Earth's magnetic field

The magnetic field of the Earth resembles that of a giant magnet in the center of the Earth, pointing north and south. A compass (below) has a magnetized needle pivoted at its center so that it can swing from side to side. The right-hand side of the diagram (below right) shows how a compass needle lines itself up with the magnetic field. As a result, it always points to the north. The magnetic and geographic poles are usually in slightly different places. A dip needle is a magnetic needle pivoted so that it can swing up and down. It also follows the magnetic field lines, and points straight up or straight down when over the poles.

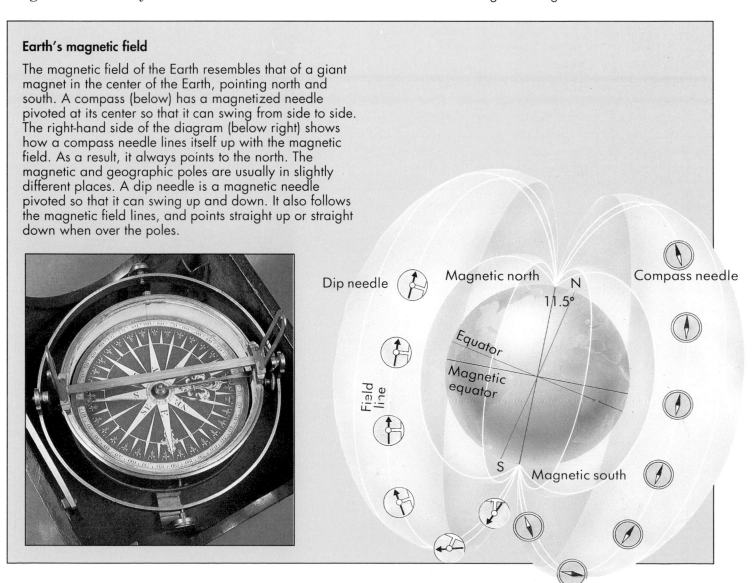

Earth's motion

At one stage when the Solar System was forming, the planets were nothing more than a series of rings of loose particles around the early Sun. It was gravitational force that held the particles in each ring. They were being pulled inward toward the Sun, but at the same time they were moving along so fast that they were flying past it. At a certain speed the two movements balanced and the particles fell into orbit round the Sun, never reaching it but never breaking away. An orbit is a circular or elliptical path in which the tendency of a body to fly away into space is just balanced by the force of gravity pulling it inward.

As the rings solidified into planets, they remained in orbit around the Sun. Many planets have smaller bodies in orbit around them. These are satellites or moons.

The Earth's orbit is not circular, but is slightly elliptical. At its closest point to the Sun (perihelion) it comes to within 147,100,000 km (91,400,000 mi.). At its farthest point (aphelion) it is 152,100,000 km (94,500,000 mi.) away. The Earth reaches perihelion in early January, and aphelion in July. It takes $365\frac{1}{4}$ days for the Earth to travel once round its orbit. This is the period of time we call a year.

As the Earth travels in its elliptical orbit, it also spins, or rotates, on its own axis. It does this every 24 hours, giving us our days and nights. The Earth's axis is tilted at an angle ($23\frac{1}{2}°$) to the plane of its orbit round the Sun. At perihelion the North Pole is tilted away from the Sun and the South Pole tilted towards it. At aphelion the tilt is reversed, the South Pole tilting away from the Sun and the North Pole tilting towards it. It is the tilt that gives rise to the change of seasons throughout the world.

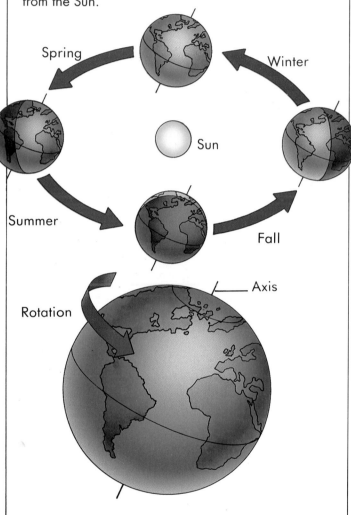

Changing seasons

It is the tilt of the Earth's axis that creates the various seasons. The Earth's axis is tilted at $23\frac{1}{2}°$ to the plane of its orbit around the Sun. As the North Pole tilts toward the Sun in June, the Sun does not set at all in the far north, and it is northern summer. It is northern winter when the North Pole tilts away from the Sun.

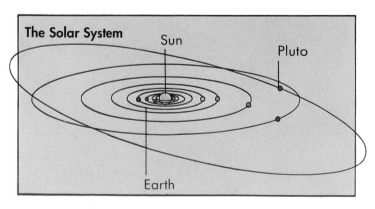

The Solar System

◀ The orbits of nearly all the planets in the Solar System lie in much the same plane. This would have been the plane of the original disk of gas that formed the Solar System. A notable exception is Pluto, which has an orbit that is tilted in relation to this plane. Pluto's orbit is also highly elliptical, and at its perihelion the planet actually lies inside Neptune's orbit.

12

▼ When seen from space, the Earth looks perfectly round. The deepest ocean basins and the highest mountains hardly blemish its spherical surface. But the Earth is slightly flattened at the North and South Poles. As the Earth spins on its axis, the rotation makes it bulge outward along the Equator.

Structure of the Earth

As the Earth solidified from its nebula of gas and dust, its components separated themselves according to their densities. Perhaps the densest substances congregated first, and then the less dense ones gathered on the outside. This would have resulted in a structure with a heavy central core and a lighter covering. Alternatively, all the substances may have clumped together at the same time. Then the densest of them would have sunk toward the center and collected there. Whatever happened, we now have a planet with a massive central core covered by a number of different layers.

At the center lies the core. It is made of the heavy metals iron and nickel. There are two layers here. The inner core is solid, whereas the outer core is liquid.

Around the core lies the rocky mantle, which also consists of two layers. The mantle makes up the bulk of the Earth. It is mostly solid and is made of silicates, compounds of silicon and oxygen. Most of the Earth's rocks are made up of silicates.

On the outside is the crust. This is, to us, the most important of the layers, and it is the only one we can reach directly. There are two types of crust, made of slightly different silicate materials. The larger area consists of oceanic crust, which is quite thin. It is made largely of silica and magnesium, and is given the shorthand name "sima." The second type of crust forms the continents. It is made mostly of silica and aluminum, and is called "sial."

Sial is lighter and thicker than sima. The continents are formed of separate lumps of sial "floating" in the sima of the ocean floor. Unlike the sima, the sial cannot be carried downward because it is far less dense. As a result, the continents are much older than the ocean floors.

The layers of the lithosphere and asthenosphere, which make up and move the plates of the Earth's surface, are found toward the outside of the globe. The solid lithosphere consists of the crust and the uppermost part of the mantle. The soft asthenosphere is a distinct layer of mantle positioned just below it.

▼ Pillow lavas are found on the seabed. They are cushion-shaped lumps of volcanic material formed when lava erupts from a submerged ocean ridge and cools quickly. The whole of the deep ocean floor, beneath the sediments, is formed of pillow lavas. They are part of the sima – the material of the oceanic crust. Sima is rarely seen because it lies beneath several miles of ocean water and sediments.

▼ All the minerals and rocks of mountains and plains are components of the sial. The sial – the material of the continental crust – is all around us. It tends to be more complex than the sima, because it is much older. Its rocks are constantly being lifted up by mountain-building activity and worn away by weather and rivers. The fragments are redeposited as sediments, and eventually they are turned into solid rocks once more.

14

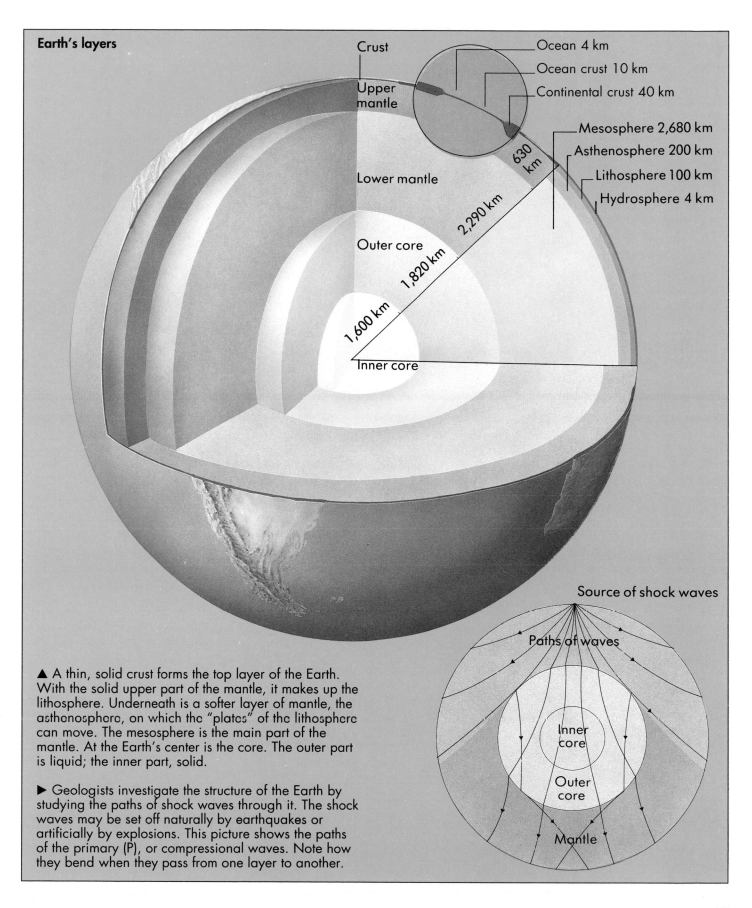

Earth's layers

Crust

Upper mantle

Lower mantle

Outer core

Inner core

Ocean 4 km
Ocean crust 10 km
Continental crust 40 km

Mesosphere 2,680 km
Asthenosphere 200 km
Lithosphere 100 km
Hydrosphere 4 km

630 km
2,290 km
1,820 km
1,600 km

▲ A thin, solid crust forms the top layer of the Earth. With the solid upper part of the mantle, it makes up the lithosphere. Underneath is a softer layer of mantle, the asthenosphere, on which the "plates" of the lithosphere can move. The mesosphere is the main part of the mantle. At the Earth's center is the core. The outer part is liquid; the inner part, solid.

▶ Geologists investigate the structure of the Earth by studying the paths of shock waves through it. The shock waves may be set off naturally by earthquakes or artificially by explosions. This picture shows the paths of the primary (P), or compressional waves. Note how they bend when they pass from one layer to another.

Source of shock waves

Paths of waves

Inner core

Outer core

Mantle

15

The skin of the Earth

The Earth's surface is in constant motion. It is made up of a series of slabs, or "plates," which shift and drift so slowly that the movement can hardly be detected. Yet after millions of years the result can be seen as continental drift – the gradual movement of the continents.

The moving surface layer of the Earth is a stiff skin, the lithosphere. It lies on top of a softer layer called the asthenosphere. The lithosphere is like a cooling scum on the more plastic asthenosphere.

▶ A cross section of the Earth shows weak points where material from the asthenosphere pushes to the surface to form new lithosphere. They occur on ridges along the ocean floor, as in the Indian Ocean Rise (1).

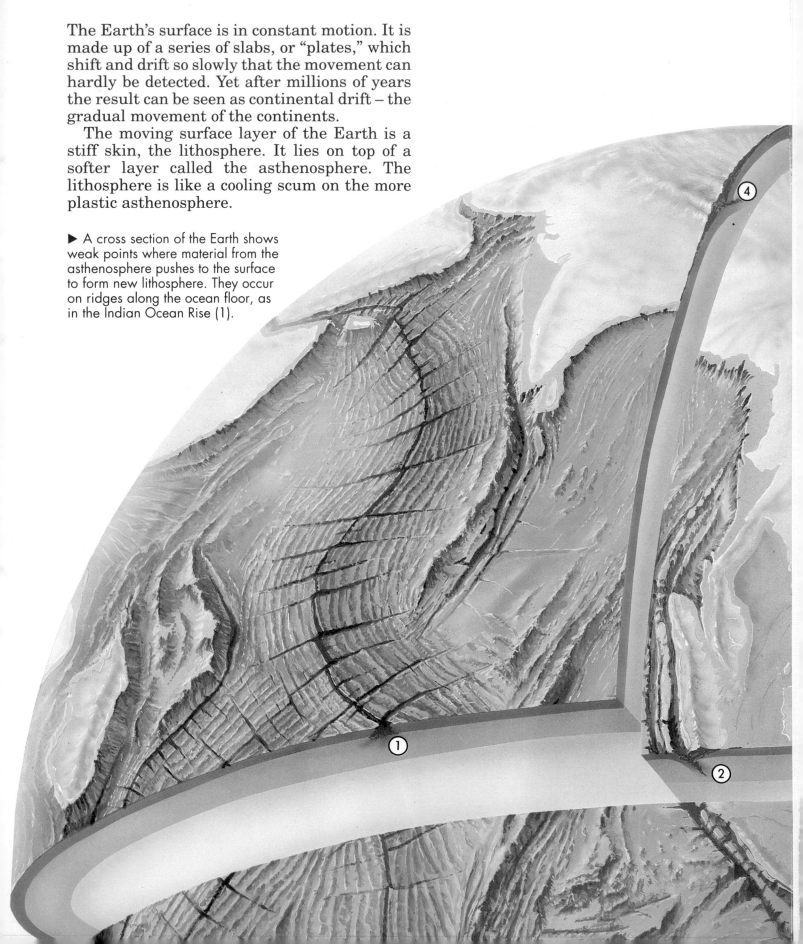

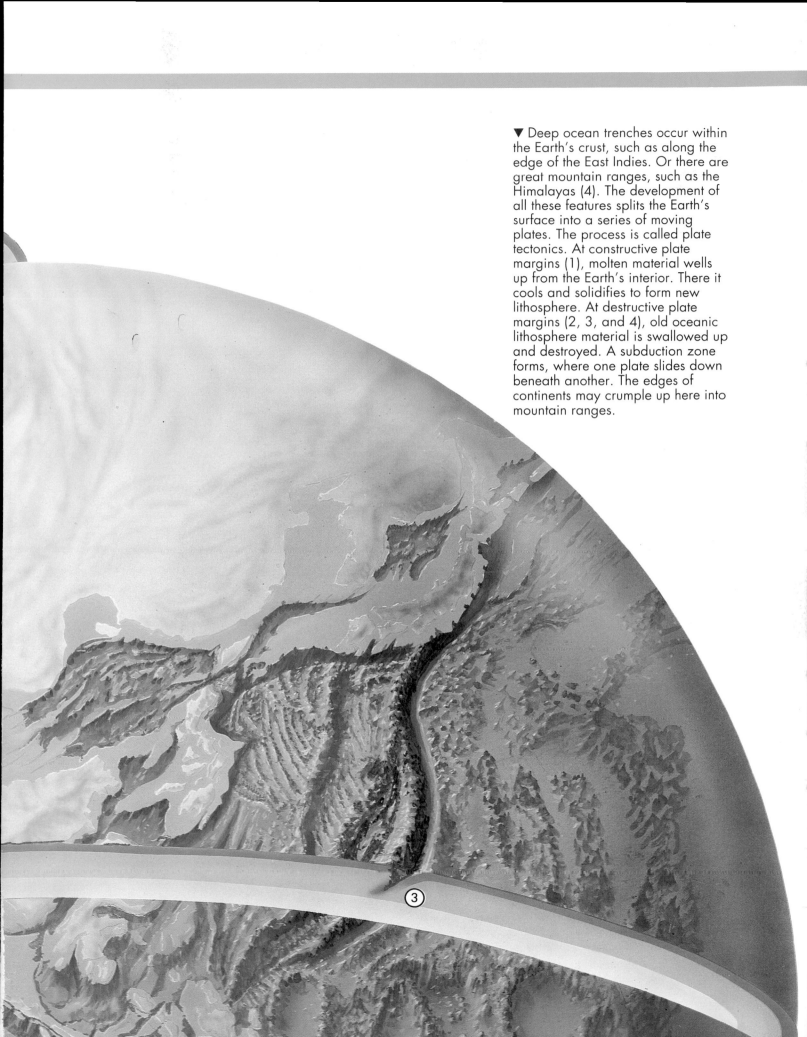

▼ Deep ocean trenches occur within the Earth's crust, such as along the edge of the East Indies. Or there are great mountain ranges, such as the Himalayas (4). The development of all these features splits the Earth's surface into a series of moving plates. The process is called plate tectonics. At constructive plate margins (1), molten material wells up from the Earth's interior. There it cools and solidifies to form new lithosphere. At destructive plate margins (2, 3, and 4), old oceanic lithosphere material is swallowed up and destroyed. A subduction zone forms, where one plate slides down beneath another. The edges of continents may crumple up here into mountain ranges.

Continents adrift

The surface of the Earth is an unstable, shifting place. Since the mid-1960s it has been known that the crust and the topmost layer of the mantle form a number of distinct plates that cover the Earth like the panels of a soccer ball. The plates are being generated continuously along one edge and destroyed along another. The shifting of continents across the globe, something suspected for centuries, is the result of all this movement. Earthquakes and volcanic eruptions are two destructive side effects that occur at plate boundaries.

SPOT FACTS

• The longest mountain range is the Andes in South America. It is 8,900 km (5,500 mi.) long.

• The highest mountain is Mount Everest in the Himalayas at 8,848 m (29,028 ft.).

• The biggest continent is Eurasia – Europe and Asia combined – with a land area of 54,750,000 sq km (21,140,000 sq. mi.).

• Seafloor spreading occurs at about the same rate as the growth of fingernails.

• The Atlantic Ocean is roughly 10 m (33 ft.) wider today than it was when Christopher Columbus sailed to America in 1492.

• Some 50 million years ago, Australia was joined to Antarctica.

• The Atlantic Ocean is widening at the expense of the Pacific. In the past 130 million years, an area of oceanic crust equal to the size of the Pacific Ocean may have been thrust under North and South America.

• As early as 1620, the philosopher Francis Bacon remarked on the fact that the eastern and western shores of the Atlantic follow a similar space, like a jigsaw puzzle.

• The opposite shores of the Atlantic Ocean were once joined together. Rocks in Brazil and West Africa match in age, type, and structure; and they often contain fossils of creatures that could never have swum from one continent to the other.

• At the San Andreas fault in California, two tectonic plates are sliding past each other. This has caused many earthquakes in the past. The most famous, which occurred in 1906, destroyed much of downtown San Francisco.

• In 1976, the city of Tangshan in China was destroyed by a powerful earthquake measuring 7.9 on the Richter Scale. The earthquake killed 750,000 people, but a group of 10,000 coal miners who were underground at the time escaped unharmed.

• In Iceland, the Mid-Atlantic ridge rises above sea level. There are many volcanoes and lava flows there, as well as geysers and boiling mud lakes.

• The Himalayas were formed when India, moving north, ran into the stationary Asian continent. Although the main uplift occurred 12 to 65 million years ago, the process is still going on.

• Volcanoes release huge amounts of energy. Krakatoa in Indonesia erupted in 1883 with a force similar to the detonation of a billion metric tons of TNT. Waves 16 m (50 ft.) high were produced, and traveled up to 12,000 km (8,000 mi.). The blast produced the loudest noise in history; one explosion was heard 4,830 km (3,000 mi.) away.

• Subduction of oceanic crust, with the associated earthquakes and volcanoes, give the rim of the Pacific basin the name "Ring of Fire."

• In 50 million years' time, Australia will probably begin to collide with Asia, a new ocean may be opening in East Africa, and Los Angeles will have been carried north of San Francisco's current position.

Dynamic Earth

The greatest mountain ranges on Earth – the Himalayas, Rockies, Andes, Alps, and Urals – were formed by a folding action of the upper layers of the Earth's crust. As a result, the rocks are twisted and squeezed, obviously deformed under some great pressure. It is the movement of the Earth's surface plates – the action of plate tectonics – that has brought this about.

Each plate is created from molten material from the Earth's interior. This wells up in volcanic activity along the great ridges that run along the ocean beds. The plates build out from their edges and move apart. Eventually, when a plate meets another plate traveling in the opposite direction, the edge of one plate slides beneath the other and is destroyed.

Sometimes this destructive plate margin lies along the edge of a continent, as it does along the western coast of South America. Then the edge of the continent is crumpled up into a mountain range – in this instance the Andes. Often two continents collide and produce a range where the two continents have fused. The Himalayas and Urals formed in this way.

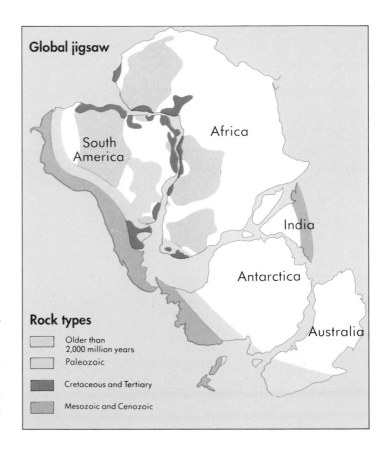

Global jigsaw

Rock types
- Older than 2,000 million years
- Paleozoic
- Cretaceous and Tertiary
- Mesozoic and Cenozoic

▲ Continents may split apart as new constructive plate margins develop beneath them. South America, Africa, India, Antarctica, and Australia were once a single landmass known as Gondwanaland. They broke up and drifted apart as new oceanic crust developed between them. We know this because of the shapes of the continental shelves, the similarity of fossils on each landmass, the continuation of mountain ranges across them, and similarity of rock types.

◀ The intensely folded rocks of Wales are the result of past movements. Once there was another ocean where the Atlantic now lies. This closed up as the landmasses on the east and west collided. They threw up a mountain range like the Himalayas. After many millions of years the continents broke apart again and the modern Atlantic Ocean was formed. The remains of the old mountain range lie in the Appalachians of North America, and in the highlands of Wales, Scotland, and Norway in Europe.

Surface plates

In the 1960s, scientists studying the bottom of the Atlantic Ocean began to notice something that led to a revolution in geological knowledge. All rocks contain magnetic particles, and when they are formed, their magnetism lines up along the Earth's magnetic field. At the crest of the Mid-Atlantic Ridge, the ocean ridge that runs north to south in the Atlantic Ocean, the magnetism of the rocks lines up as expected. But to each side the rocks have a reversed magnetism, evidently produced at a time of magnetic reversal.

These magnetic reversals happen from time to time, when the Earth's magnetic North Pole and South Pole change polarity. Further exploration found that down the sides of the ridge the rocks showed alternating stripes of normal and reversed magnetism. The pattern on one side was the mirror image of the pattern on the other. This indicates that the seabed is being created along the crest of the ridge, and is moving away to each side as new material continues to well up from below. As each band of new rock is formed, it lines up with the direction of the Earth's magnetic field at the time. This process is called seafloor spreading.

Further proof that the rocks of the seabed are spreading came when it was noticed that the rocks on the ridge of the crest are exposed, whereas those farther away are covered in sediment. The sediment layer becomes thicker even farther from the crest, reflecting how much older the seabed must be.

Soon this new concept of seafloor spreading was put together with the older discovery of continental drift to produce the new science called plate tectonics.

The changing face of the Earth

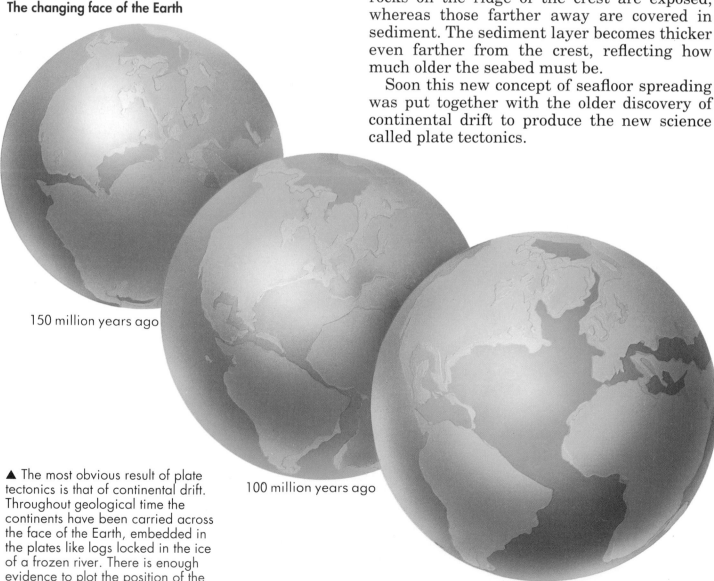

150 million years ago

100 million years ago

50 million years ago

▲ The most obvious result of plate tectonics is that of continental drift. Throughout geological time the continents have been carried across the face of the Earth, embedded in the plates like logs locked in the ice of a frozen river. There is enough evidence to plot the position of the continents at different stages of the Earth's history.

20

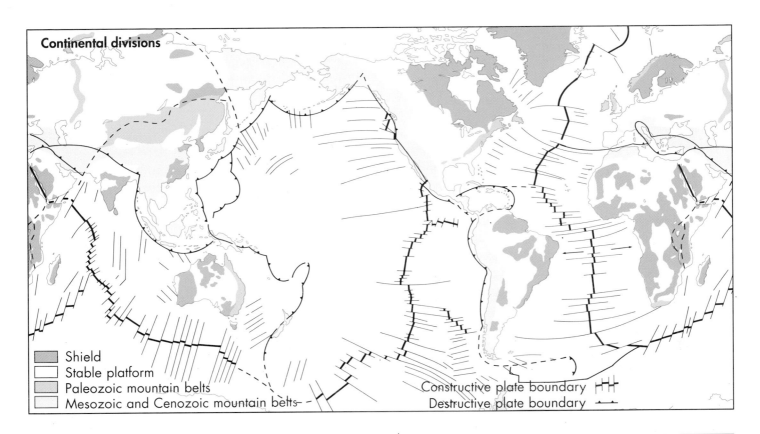

Continental divisions

Shield
Stable platform
Paleozoic mountain belts
Mesozoic and Cenozoic mountain belts

Constructive plate boundary
Destructive plate boundary

▲ Continents are made up of three basic components: the shield, a flat plain of ancient rock; the stable platforms, the base of which is made up of the same kind of rock as the shield, but with a layer of different rock on top; and the mountain belts. The oldest mountains – the Paleozoic – are nearest the shield, and the youngest – the Mesozoic and Cenozoic – are farthest away from it.

Today

Wegener the pioneer

In 1912, the German meteorologist Alfred Wegener (1880–1930) came up with the idea of continental drift based on sound scientific reasoning. He produced a series of maps of the world as it was in the past. They were essentially the same as those that can be produced today with our vastly increased knowledge. However, he could not account for the mechanics of the movement, and did not live to see plate tectonics provide the answer.

Rifts and mountains

All the large-scale features of the Earth's surface, including its highlands and lowlands, mountain ranges and plains, can be looked at as the result of the activity of plate tectonics.

The most extensive mountain ranges are those of the fold mountains. They are caused by compression – by two plates grinding into each other. When two plates, topped with oceanic crust, meet each other, one is "subducted." It slides down into the depths of the mantle, pulling down the seabed into an ocean trench, while the other plate rides up above it. As the descending plate melts, the molten material rises up through the overlying plate and bursts through to the surface as a series of volcanoes. These grow until they rise above the surface of the sea as an arc of volcanic islands. Many such island arcs, running parallel to ocean trenches, are found along the northern and western fringes of the Pacific Ocean.

A slab of continental crust may be embedded in the oceanic crust of the plate that is being destroyed. Eventually this continent reaches the ocean trench at the subduction zone, and there the movement stops. Continental crust is too light to be drawn down into the mantle.

If the plate movement continues, the oceanic crust of the opposite plate then begins to slide down beneath the continent. This establishes another subduction zone and ocean trench. The edge of the continent deforms and crumples up with the activity, while sediments from the descending plate are scraped off against it. Molten material from below thrusts up through the continental edge and a series of volcanoes develops in the deformed margin. As a result, a

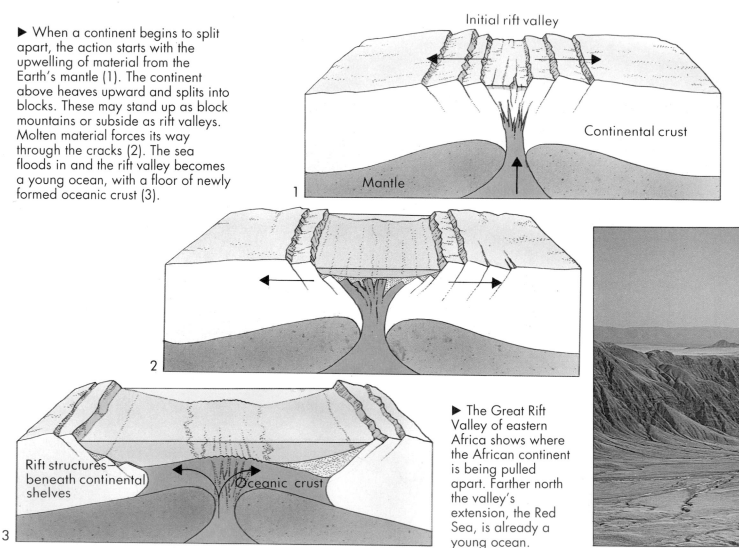

▶ When a continent begins to split apart, the action starts with the upwelling of material from the Earth's mantle (1). The continent above heaves upward and splits into blocks. These may stand up as block mountains or subside as rift valleys. Molten material forces its way through the cracks (2). The sea floods in and the rift valley becomes a young ocean, with a floor of newly formed oceanic crust (3).

▶ The Great Rift Valley of eastern Africa shows where the African continent is being pulled apart. Farther north the valley's extension, the Red Sea, is already a young ocean.

22

very complex chain of fold mountains grows along the continental edge. The Andes, along the western edge of South America, are a particularly good example, with an ocean trench just offshore. There are also a great number of volcanoes along their length, which is another typical feature of fold mountains.

The subducting plate may bring its own continent, and when the two collide the resulting mountain range is particularly enormous, such as the Himalayas between India and Asia. Now the movement really does stop. But the forces are still at work, and make themselves felt at some other part of the globe. A new constructive plate margin develops somewhere. If it is in the middle of a continent, the continent heaves up and cracks to produce the other major type of mountain range, the so-called block mountains.

As a result, all the continents have the same general pattern. There is a central area of hard old rock, usually worn flat with age, surrounded by successively younger ranges of fold mountains. A rift valley among block mountains may show where the continent is being pulled apart. If one border of the continent shows the cracks and block mountains that we would associate with a rift valley, then the continent has probably broken away from another one some time in the past.

Coastal volcanoes

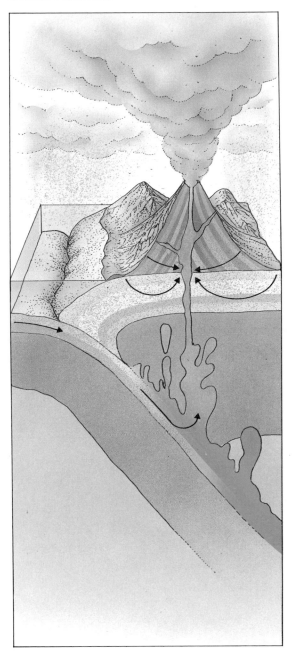

When one plate slides below the next, it is eventually destroyed in the mantle. Friction between the two plates melts the rock along the boundary, and this rises through the plate above, eventually forming volcanoes on the surface. The melting of the rocks is helped by the presence of seawater brought down by the moving plate, and a lot of water is erupted from the volcanoes.

23

Volcanoes

Volcanoes occur where hot material leaks out from the Earth's interior. This usually happens at the margins of the tectonic plates.

At constructive plate margins, along the mid-ocean ridges, the hot material from the mantle wells up and solidifies. The molten matter that is forced out at the surface is called lava. It solidifies not far from the vent, gradually building up into a mountain. The activity takes place on the ocean floor, and so these types of volcanoes are rarely seen. Only in places such as Iceland does the mid-ocean ridge reach above the surface of the water. Then the volcanoes can be seen on land.

At destructive plate margins the molten material comes from the breakdown of the plates themselves, and the volcanoes form in island arcs or in fold mountains. The lava is a different type from constructive margin lava, and forms different types of volcanoes.

A third type of volcano is found away from the plate margin, over a "hot-spot" of activity deep in the mantle. The lava that erupts is of the same type as that found at a constructive plate margin, and the same kind of volcanoes are thereby produced.

Volcanoes and earthquakes

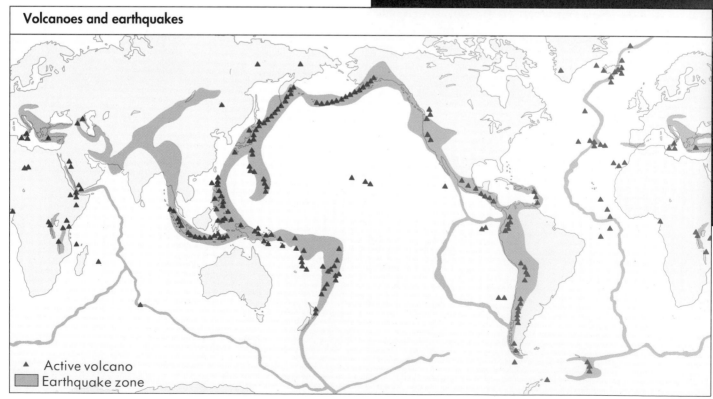

▲ Active volcano
▨ Earthquake zone

24

▲ A river of lava pours out from an erupting volcano. It is formed from magma from the mantle. As the magma rises and cools, some of its minerals solidify and sink back. Gases are given off as bubbles. The resulting lava does not have the same composition as the mantle.

◄ Most of the world's volcanoes are found along the margins of the plates. This distribution, as well as the distribution of earthquakes, shows where most of the plate tectonic activity is taking place.

Types of volcanoes

At destructive plate margins the lava is rich in silica. This makes it stiff, producing steep-sided volcanoes such as the composite volcanoes and cinder cones. Eruptions can be very violent, and the rock formed is called andesite. At constructive plate margins the lava contains less silica. It is runny and erupts quietly, producing broad shield volcanoes and fissures. The rocks that form are called basalts.

Composite volcano

Cinder cone

Shield volcano

Fissure

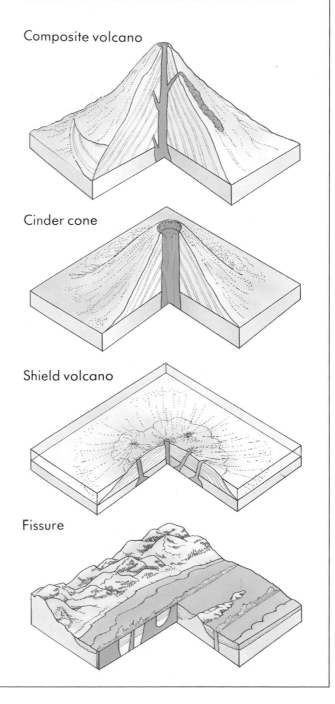

Earthquakes

Through the action of plate tectonics the crust of our planet is always moving. Over millions of years the continents drift from one place to another. They jostle together and push up mountain ranges. The movements do not take place continuously, but in small jerks and jumps. It is these jumps that set up the vibrations we call earthquakes.

The forces that sometimes move the crust of the Earth work all the time, and cause stresses to be built up in the rocks. Eventually the stresses become so strong that they make the rocks snap. The rocks whip along a crack, called a fault, and this movement causes an earthquake shock. The shock waves travel outward from the focus, the point where most of the movement takes place, like the ripples from a splash in a pond. The point on the Earth's surface directly above the focus is called the epicenter. Most damage is usually done there.

When the pieces of the Earth's crust snap along a fault, they usually move too far. Later they may spring back some distance and produce aftershocks, and this may continue until the rock masses have settled. Then the stresses begin to build up once more until they are released by the next earthquake.

The different types of shock wave produced by an earthquake travel at different speeds. Geological observatories around the world can detect them with instruments called seismographs. By timing when waves arrive, scientists can tell how far away the earthquake was. A world network of observatories can now pinpoint any earthquake's focus.

The places most likely to have earthquakes are the edges of the Earth's surface plates, where the plates are jostling each other and being created or destroyed. But earthquakes cannot be predicted.

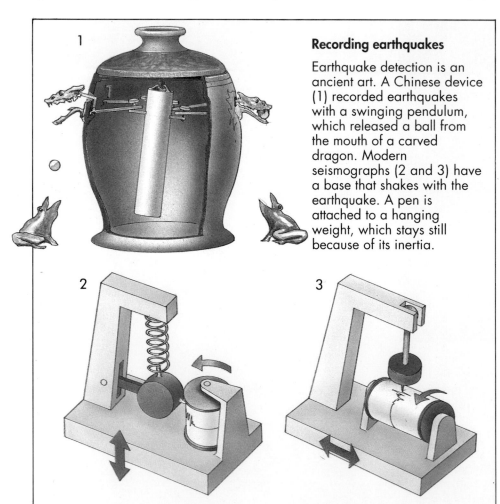

Recording earthquakes

Earthquake detection is an ancient art. A Chinese device (1) recorded earthquakes with a swinging pendulum, which released a ball from the mouth of a carved dragon. Modern seismographs (2 and 3) have a base that shakes with the earthquake. A pen is attached to a hanging weight, which stays still because of its inertia.

▶ The result of a severe earthquake that struck Mexico City in September 1985. A multistory building has collapsed like a pack of cards. During this earthquake, more than 10,000 people died.

▼ The notorious San Andreas fault near San Bernardino in California. Movements along the fault cause frequent earthquakes.

Rocks and minerals

Several kinds of rocks make up the Earth's crust. They have formed in different ways: some from red-hot lava that spewed out of volcanoes, some from rock debris swept down from the mountains by rushing water, and some from the fossils of ancient sea creatures. All the rocks are made up of collections of minerals, usually in the form of glassy crystals packed haphazardly together. But here and there, in cavities in the rocks, the crystals have room to grow into beautiful shapes, "the flowers of the mineral kingdom."

SPOT FACTS

• The most common elements in the Earth's crust are oxygen and silicon, usually united as silica. Together they make up just over three-quarters of the crust.

• Although 87 elements occur in minerals, eight of them make up 99 percent of the rocks of the Earth's crust. The oldest rocks are 3,800 million years old, and are found in Greenland.

• Chalk is formed from the microscopic remains of tiny creatures that lived in the seas during the Cretaceous period of Earth's history, which was between 144 million and 65 million years ago. "Cretaceous" means chalky.

• The rocks at the top of Mount Everest were formed at the bottom of the sea some 50 million years ago.

• Igneous rocks are usually quite different in composition from the mantle from which they were formed – except in the case of the Troodos Mountains of Cyprus, where there are rocks thought to be composed of solidified mantle.

• Lava cools the fastest of any igneous rock. It can do so in midair during an eruption; the sound of lava landing resembles that of breaking crockery.

• Large, floating lava islands were a shipping hazard for a year after the eruption of Krakatoa in 1883.

• Diamond, the hardest and one of the most precious minerals, is a form of carbon, which also forms coal, graphite, and soot. Industrial diamonds are widely used for cutting, grinding, and polishing other hard materials.

• Unusual sedimentary rocks called varves show annual layers like the growth-rings of a tree. This is caused by seasonal changes in the way the sediment is laid down.

• Sediments of the ocean floor are laid down in most areas at about 1 mm (0.039 in.) every 1,000 years.

• Included in deep-ocean deposits are very fine clays, dust from meteorites, sharks' teeth and whale earbones, materials from undersea volcanic eruptions, and nodules of valuable metal ores.

• Minerals can be classified according to Moh's scale of hardness. This starts at 1 with the softest, talc, and goes up to 10, which is diamond.

• The German poet Goethe (1749–1832) collected minerals in his spare time; the common ore of iron, Goethite, is named for him.

• Lapis lazuli is a beautiful and complex mixture of minerals found chiefly in the Kotcha Valley in Afghanistan. Marco Polo visited the mines in 1271, and they are still worked today.

• The red mineral realgar was for centuries used as a pigment in paints, until it was discovered to be very poisonous. Realgar is arsenic sulphide.

• The mineral magnetite (an iron oxide) can occur as natural magnets, or lodestones. These were used to magnetize needles to make the first magnetic compasses used for navigation.

• The oldest fragments of the Earth found so far are the zircon crystals discovered in 1984 in Western Australia, dated at 4,276 million years old.

Reading the rocks

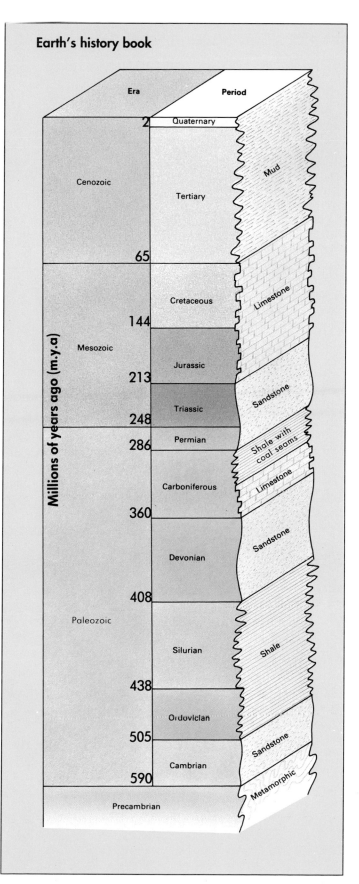

Earth's history book

The Earth's crust consists of three types of rocks. When molten material from the Earth's interior solidifies, it forms igneous rock. When fragments of sand, silt, or rubble are laid down, compressed, and cemented into a solid mass, the result is a sedimentary rock. When pre-existing rocks are crushed and "cooked" deep within a mountain range, their mineral content changes. The resulting rocks are termed metamorphic rocks.

Sedimentary rocks are most common at the Earth's surface, and by studying them we can find out much about what happened in the past. Sediments accumulate under specific conditions. So when we see a bed of sandstone with ripple marks in it, we can deduce that it was formed from a bed of sand laid down in a shallow sea. If, above it, there is a bed of mudstone containing the fossils of freshwater snails, we can deduce that a river later swept mud into the area and covered the sand.

Examples like this have enabled us to write the full history of the Earth's surface from the days when sediments first began to form.

◄ Two centuries of studying the rocks has enabled geologists to work out a time scale of the Earth's history. The sequence of rocks along the side is an imaginary one, giving a rough idea of the layers that might be found.

▼ Fossils, such as these ammonites, can help to tell the age and the history of a rock, because the animals lived only during a particular period and under certain climatic conditions.

29

Changing rocks

We just need to look around us at the conditions of the Earth's surface today to realize the range of conditions that must have existed in times past. We know that we get different kinds of sand in rivers, beaches, and deserts. We see that streams produce silt, and lakes become choked with peat. Coarse shingle, or gravel, gathers on the shoreline. At the bottom of the sea, muds, oozes, and fragments of seashell and coral accumulate. All these sediments eventually produce different sedimentary rocks.

Within these rocks there are different structures. Sand laid down in a river forms twisted beds that reflect the effects of the water current. The shapes of sand dunes can be seen in sandstones formed from desert sands. Coarse sediments form from particles deposited in strong currents, and fine sediments in gentle currents. Cracks develop in drying mud, and these can be preserved in the subsequent rocks. The study of this kind of evidence is the science of historical geology, also called stratigraphy.

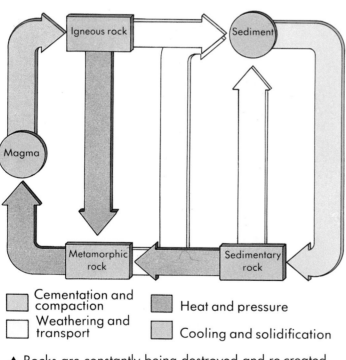

▲ Rocks are constantly being destroyed and re-created, a process known as the rock cycle. Many stages are involved. Sediments form sedimentary rock, which may later be crushed and re-formed in the heart of a mountain to produce metamorphic rock. The sedimentary rock may even melt with the heat and later solidify into igneous rock. All types may crumble when exposed, and their debris forms new sediments.

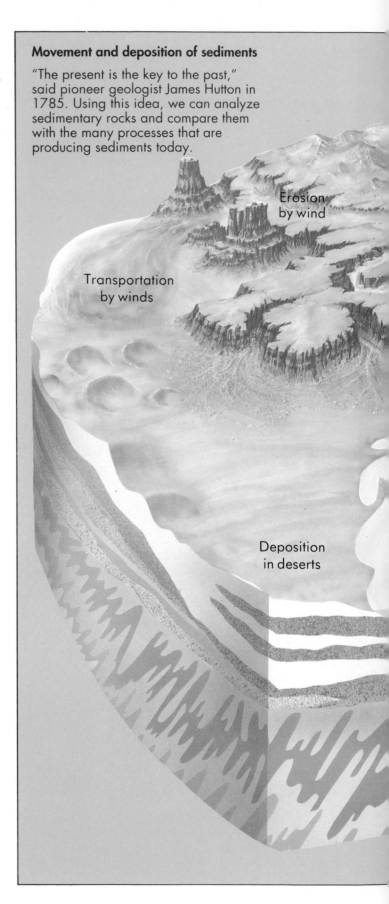

Movement and deposition of sediments

"The present is the key to the past," said pioneer geologist James Hutton in 1785. Using this idea, we can analyze sedimentary rocks and compare them with the many processes that are producing sediments today.

30

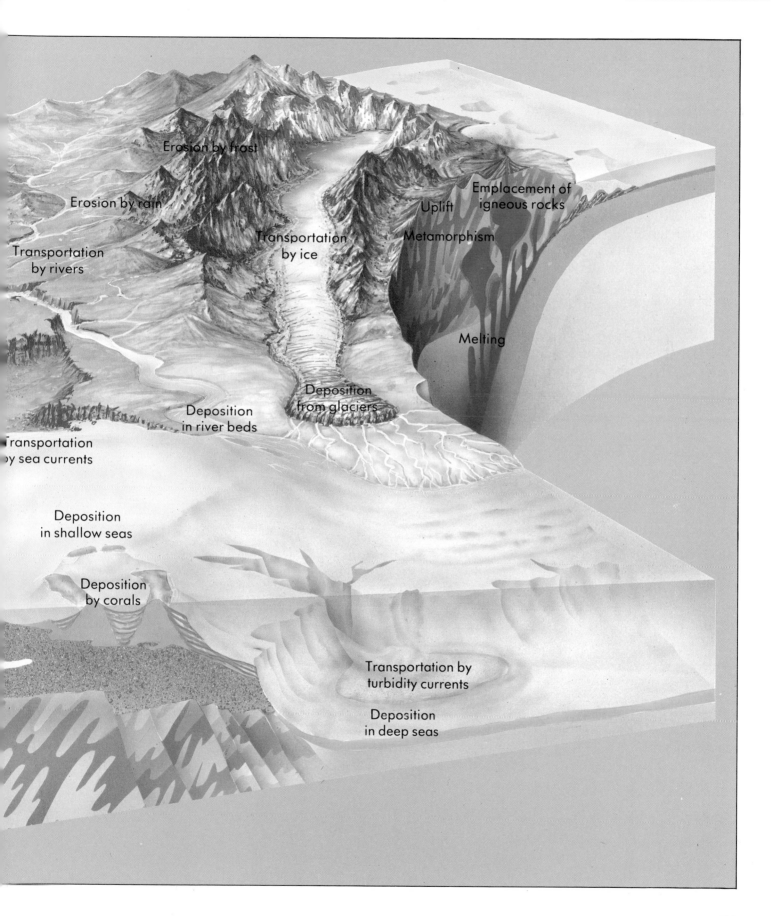

Erosion by frost

Erosion by rain

Transportation by rivers

Transportation by ice

Uplift

Metamorphism

Emplacement of igneous rocks

Melting

Transportation by sea currents

Deposition in river beds

Deposition from glaciers

Deposition in shallow seas

Deposition by corals

Transportation by turbidity currents

Deposition in deep seas

Igneous rocks

Perhaps the simplest type of rock, and the most easily understood, is igneous rock. Its formation is theoretically quite straightforward. Molten material from inside the Earth cools and becomes a solid mass.

There are two main types of igneous rock – intrusive and extrusive. Intrusive rocks form when a mass of molten material is injected into the rocks of the crust and solidifies there without reaching the surface. We see intrusive igneous rocks only when the rocks above have been eroded away. Extrusive rocks form when the molten material cools on the surface, as for example, in a lava flow.

Intrusive rocks cool very slowly, and so they tend to be coarse-grained. The crystals of individual minerals are big enough to be seen by the naked eye. Granite is a good example. Extrusive rocks cool quickly and so are fine-grained. They have microscopic mineral crystals. Basalt is an example. Sometimes the molten rock begins to cool underground and the first minerals form large crystals. Then the whole mass bursts out at the surface and solidifies quickly. The result is a rock called a porphyry, which consists of a fine "groundmass" with big crystals embedded in it.

Composition of igneous rocks

Igneous rocks are also classified by their composition. This is rarely the same as the original magma, the molten material from the Earth's interior. The magma is rich in the chemical silica and forms silicate minerals.

As the molten material rises through the Earth's crust and begins to cool, some minerals crystallize out before the others. The silicate minerals olivine and pyroxine are early crystallizers. These, which are rich in iron and magnesium, sink to the bottom of an intrusion. Silicate minerals such as feldspars and micas, which are low in iron and magnesium but rich in the lighter metals potassium and sodium, tend to solidify later. Uncombined silica forms

◀▼ The wrinkled, ropy surface is typical of basalt, a common extrusive basic igneous rock. After lava erupts from a volcano, it may flow for some distance over the ground as a river of fire. During this time its surface cools and hardens. The chilled surface is dragged along by the movement of the liquid beneath, and it twists and distorts as it goes. The ropy lava is known by its Hawaiian name of "aa." The islands of Iceland and Hawaii were built up from the seabed by successive layers of lava flows like these.

the mineral quartz. As a result, dark iron-rich rocks form deep down, whereas light-colored silica-rich rocks form closer to the surface. Geologists call the darker rocks basic and the lighter rocks acidic.

Igneous rocks that form from magma being brought up at a constructive plate margin tend to be basic. Coarse intrusive dolerite and fine extrusive basalt are formed there. At a destructive plate margin, the magma is made of molten plate material and is usually richer in silica. The rocks that form there tend to be acidic igneous rocks, such as intrusive granite and extrusive andesite.

Although the silicate minerals are rich in metals, it is difficult to remove the metals from the silica. They are therefore no good as ores. Silicate minerals are thus referred to as the rock-forming minerals. Ore minerals – used as sources of metals – are usually sulfides or oxides of metals, and they do not usually form the bulk of the rock.

▶ Sheer granite cliff faces provide a spectacular sight. El Capitan cliff in Yosemite National Park shows little sign so far of any erosion.

▼ Granite is a typical intrusive igneous rock. Its feldspar minerals break down easily on exposure to air, and so exposed granite wears away into rounded lumps.

Sedimentary rocks

The rocks that are formed when a layer of mud, sand, or other natural debris is compressed and cemented together are called sedimentary rocks. Like igneous rocks, they can be classified according to their origin. The main type is clastic sedimentary rock. This is formed when fragments of other rocks, such as shingle or sand, are compressed.

Any rock that is exposed at the surface of the Earth is worn away by the relentless onslaught of the wind and rain. Some of its minerals, such as feldspar, may be dissolved away by acid in rainwater. Or the rock may be broken apart by the expansion of ice in its cracks. All these actions make rock break down into fragments that can be washed away by streams or even blown away by the wind. Eventually the fragments settle. The coarsest fragments do not travel far, but come to rest as boulders, cobbles, and shingle at the foot of a cliff or on a shoreline. Finer pieces like sand and silt can be carried farther and deposited on beaches, in deltas, or estuaries. The finest matter is washed out to sea and settles as mud. These various sediments may eventually become sedimentary rocks such as conglomerate, sandstone, and shale.

The second classification is biogenic sedimentary rock. It is formed from fragments of once-living matter, such as corals or seashells. These form limestone in which you can often see the fossils of the creatures that originally formed the rock.

▲ Cheddar Gorge, England, is formed of limestone. This is a very common sedimentary rock that forms in distinct layers or beds, which may be either very thin or very thick.

▼ The sand in the foreground may eventually become sandstone like that in the background. This will only happen once the sands have become buried, compressed, and cemented together. The sandstone will appear at the surface only if the whole area is caught up in mountain-building activity, and the overlying beds are worn away. The beds that were once horizontal will be tilted up and twisted. They may be distorted by folds, or they may shift along cracks called faults.

Finally, chemical sedimentary rock forms when substances dissolved in the water come out of solution and form a crust on the bottom of a lake or the sea. Rock salt and certain kinds of limestones form in this way when lakes and shallow seas dry up.

Natural concrete

When sediments are buried, they become compressed beneath the weight of the sediments on top of them. Then groundwater seeps through them, depositing mineral crystals on and between the fragments. This cements them together, just like cement holds together the gravel and sand in concrete, and turns the loose material into a hard, solid mass.

▲ Some sedimentary rocks have practical uses. The coal being mined here is a biogenic sedimentary rock made from ancient vegetation. Many sandstones and limestones make good building materials.

Sedimentary rock types

We can usually identify a sedimentary rock type by looking at the fragments from which it is made up. A chemical sedimentary rock, such as rock salt (1), is made up of crystals, rather like those in an igneous rock. A clastic sedimentary rock, such as a conglomerate (2), consists of distinct lumps. The lumps may be rounded if water has worn away at them for a long time, or jagged and angular if they have not traveled far. They may also be coarse or fine. In a biogenic sedimentary rock, such as chalk (3), we can see the fragments of shells. The example shown here is of a microscopic fossil, but often the shell fragments can be seen with the naked eye.

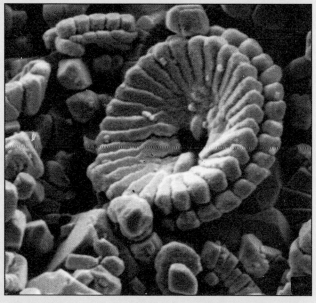

Metamorphic rocks

When the drifting continents grind into each other, the immense pressures involved crumple up mountains and alter the rocks deep inside the crust. When a rock is subjected to so much pressure and heat that its minerals change, the rock becomes a metamorphic rock. But the rock remains solid throughout this process. If it melts and then solidifies again, it changes into an igneous rock.

Geologists recognize two main types of metamorphic rock. The first is thermal metamorphic rock. This is formed principally by heat. It is quite often found in localized patches around igneous intrusions, where the heat of the intrusion has re-formed the surrounding rock. A thermal metamorphic rock can be difficult to distinguish from an igneous rock, because it tends to consist of masses of intergrown crystals with little recognizable pattern. Thermal metamorphic rocks, however, contain minerals found in no other type of rock.

The second type of metamorphic rock is formed by pressure. The roots of whole mountain chains can be altered in this way. For this reason, the rock is called regional metamorphic rock. Slate is an example. The new minerals that form may do so in contorted layers and bands, often at right angles to the direction of the pressure. The very ancient terrains in the centers of continents are usually formed from regional metamorphic rocks.

▼ A slate quarry. Slate is a fine-grained regional metamorphic rock, and one of the few that is economically valuable. The pressure that formed it produced new crystals of mica, all aligned in the same direction. Mica crystals form sheets, and for this reason slate easily splits into thin slices that can be used for covering roofs and for other purposes.

Identifying crystals

Individual minerals can be difficult to identify just by looking at them with the naked eye. Color is not a good guide, because any mineral can contain traces of an impurity that changes the color completely. Quartz, for example, can be transparent, milky white, pink, or brown. Corundum, an aluminum oxide, can be discolored red, which would make it a ruby; or it can be blue, making it a sapphire.

However, the "streak" of a mineral is quite distinctive. If a sample of mineral is scraped over a hard surface, it leaves a streak of fine powder. The color of that powder is constant for a particular mineral whatever impurity it may contain.

Crystal form is a good indicator, but usually in a rock the crystals are crammed together and show no good shape.

A useful test is hardness. Some minerals are harder than others and can be tested by scratching a sample against others of known hardness. A mineral scratches only a mineral that is softer than itself. Quartz is quite a hard mineral, and it scratches softer minerals such as calcite, but can itself be scratched only by even harder minerals such as corundum.

Luster and the way a mineral breaks, or fractures, can both be indicative. When light catches the mineral, it may have a glassy, metallic, or dull luster. When it is broken, the broken face may be straight, ragged, or shell-like.

A geologist uses a special microscope to examine a thin slice of rock and identify its minerals. The rock slice is ground until it is paper-thin and transparent. The specimen is examined using polarized light, which produces distinctive colored patterns when it passes through a mineral.

Key
1 Pyrites, with a metallic luster.
2 Flint, with a conchoidal, or shell-like, fracture.
3 Rock salt, with good crystal shape.
4 Quartz, with pink impurities.
5 Diamond, the hardest mineral of all.

Minerals may be identified by their luster, by the way they fracture, by the shapes of their crystals, or by their hardness.

37

The changing landscape

"As old as the hills" is an expression that we use when we think of something as being very old indeed. Yet compared with the age of the Earth, the hills may not be so very ancient. The Himalayas, the greatest mountain chain on Earth, are less than 50 million years old, and that is not a great span of time in geological terms. Whenever a rock becomes exposed at the surface of the Earth, and whenever an area of land rises above sea level, natural forces begin to destroy it. Gravity, running water, wind, rain, sea, and frost all work together to erode, or wear down, the landscape back to sea level. The face of the Earth is also changing because of human activity.

SPOT FACTS

- The world's highest mountains are also its youngest. The older ones have been worn down over time.

- Sea waves wear away the rocks in cliffs. During a storm, waves exert a pressure of up to 30 metric tons per sq m (a little over 1 sq. yd.).

- Erosion of the Grand Canyon began in Miocene times millions of years ago. Today it is up to 2,000 m (well over 1 mi.) deep.

- The Amazon River carries water and sediment away from an area of more than 7 million sq km (2.7 million sq. mi.).

- When water fills cracks and freezes, it expands by more than 9 percent of its volume and exerts a bursting pressure of about 150 kg per sq cm (about 330 lb. per 0.155 sq. in.).

- Granites are prone not only to chemical weathering, as the minerals break down, but also to physical weathering in the form of exfoliation. In this process, large sheets of rock come away like onion scales, such as in the Yosemite Valley in California.

- Rain is more harmful to rocks and buildings if the air is polluted. This is because crystals of various salts in the pollutants grow in the pore spaces of exposed rocks, forcing them apart.

- The rate of chemical weathering in the humid tropics is thought to be 20–40 times faster than in temperate latitudes, and is slowest in arid areas.

- It is thought that rivers have brought about a lowering of the world's landmasses by 1 m (3.25 ft.) over the last 30,000 years.

- Fast-moving streams can cut downward at rates as high as 50 cm (20 in.) in 30,000 years.

- In one year, the world's rivers transport 8 billion metric tons of materials to the sea.

- Lake Mead, behind the Hoover Dam on the Colorado River, was formed in 1936. It is one of the largest artificial lakes in the world. But if left to itself, it will fill up with sediment in the next 300 years.

- Before hydroelectric power reduced the water flow over the Niagara Falls, the Falls were receding upstream at 0.9 m (3 ft.) per year.

- On an average slope, water flows 100,000 times faster than ice. To drain an icefield, therefore, a glacier has to be far larger in cross-section than the equivalent river.

- Toward its center, the Greenland ice-sheet has been shown to be 3,000 m (9,840 ft.) thick.

- Glaciers usually move about 1 m (3 ft.) in a day. However, research at the Black Rapids Glacier in Alaska, during 1936–37, showed that it was moving more than 30 m (about 100 ft.) per day.

- Glaciers can cut valleys to below sea level. The Great Slave Lake in Canada's Northwest Territories has a floor that is 464 m (1,522 ft.) below sea level.

Weathering

Much of the debris that falls into rivers and gets washed away has been broken from exposed rocks. This has been done by the erosive action of winds, rain, flowing water, and ice.

Rainwater, even without the addition of pollution, is quite acidic. As moisture condenses from the clouds and falls in the form of drops, it dissolves carbon dioxide gas from the atmosphere. It then becomes weak carbonic acid. On soaking into the ground, the acid reacts with particular minerals in the rocks. This is very noticeable in granite areas, where the feldspars in the granite are attacked and turned into soft clay minerals. The other minerals in the granite – quartz and mica – then fall loose and are carried away as sand. This is why many granite areas have china-clay (kaolin) pits and dazzling white quartz beaches.

Another type of rock mainly affected by the acidity of rainwater is limestone. Limestone is composed almost entirely of the mineral calcite, which dissolves in weak acids. The water flowing through the rock dissolves it to form caves. Within them, the water redeposits the dissolved calcite as stalactites and stalagmites. The problem known as hard water can be due to dissolved calcite, which may be redeposited in kettles and water pipes.

The rocks of dry areas are also hugely affected by the weather. The wind in the desert can pick up sand particles and hurl them against exposed rocks and cliff faces. This wears them away in a natural "sandblasting" effect, and produces yet more sand that can erode more rock. The infrequent rains that fall in the desert soak into the surface layers of exposed rocks. This breaks down the minerals near the surface, and weakens the outermost skin. During the hot days and cold nights these surface layers expand and contract, and eventually split away to produce what is called onion-skin weathering. The rock comes apart layer by layer and leaves the core in the form of a rounded hill called an inselberg or island mountain.

Features of a cave

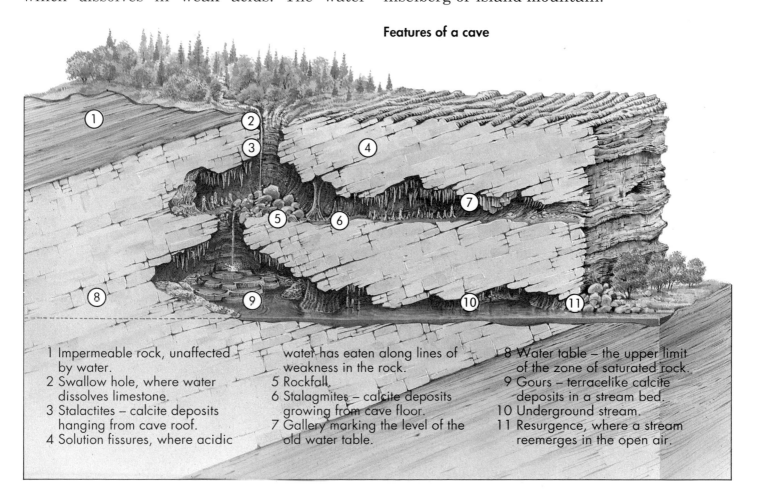

1 Impermeable rock, unaffected by water.
2 Swallow hole, where water dissolves limestone.
3 Stalactites – calcite deposits hanging from cave roof.
4 Solution fissures, where acidic water has eaten along lines of weakness in the rock.
5 Rockfall.
6 Stalagmites – calcite deposits growing from cave floor.
7 Gallery marking the level of the old water table.
8 Water table – the upper limit of the zone of saturated rock.
9 Gours – terracelike calcite deposits in a stream bed.
10 Underground stream.
11 Resurgence, where a stream reemerges in the open air.

River erosion and deposition

Much of the carving up of the landscape is done by water, and particularly by rivers. Rainwater that soaks into the ground often returns to the surface as a spring. From the spring it runs downhill as a stream.

In this young stage of a river, the water runs quickly. The faster a river flows, the more erosive power it has. It tends to erode away its bed, cutting a deep V-shaped gully as it goes. Rocks and stones picked up by the moving water are bounced along the stream bed, adding to the erosive force. Waterfalls and rapids are common at this stage.

When it leaves the hilly areas and begins to flow down a more gentle slope, the river reaches what is known as its mature stage. It still erodes the landscape, but it also deposits some of the material that it has been carrying downward. River valleys tend to be broad at this stage, much broader than the river itself. Over the years, the river's course moves about the valley floor. As a river flows round a curve, it moves more quickly on the outside. The bank at this side is undercut and worn back. On the inside of the curve the current is slower and the debris that has been carried tends to be deposited there as a beach.

The final stage of the river can be thought of as its old age. There is no valley and the water has no power to erode. It moves slowly across a plain, depositing material as it goes. Eventually it reaches the sea.

▲ A typical feature of a mountain stream is white, foaming water. This is the youthful stage of a river, when it runs fastest and is at its most violent. It bounces over rapids and waterfalls, scouring out its narrow bed. The fast-flowing river picks up rocks and gravel and carries them along toward the lowlands. By the time it reaches its mouth it is moving slowly.

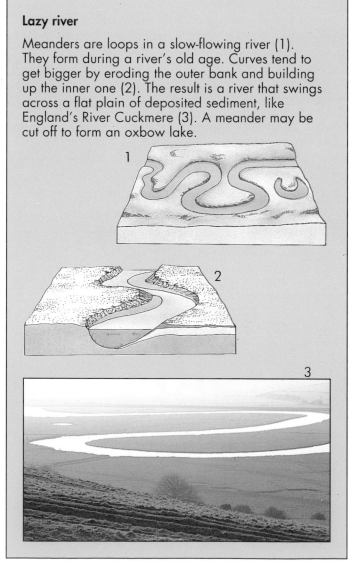

Lazy river

Meanders are loops in a slow-flowing river (1). They form during a river's old age. Curves tend to get bigger by eroding the outer bank and building up the inner one (2). The result is a river that swings across a flat plain of deposited sediment, like England's River Cuckmere (3). A meander may be cut off to form an oxbow lake.

Deltas

A river ends at its mouth. If there are few currents at that point, the debris builds up to form a delta. The delta can build out as banks along a series of channels, as in the Mississippi River (above and 1). Or it can spread out to form a semicircular area of triangular islands, as in the deltas of the Nile River (2) and the Niger River (3).

1 Mississippi

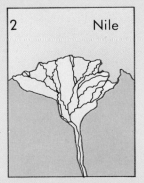

2 Nile

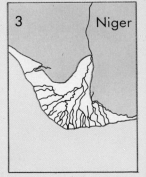

3 Niger

Ice and frost

An ever-present problem in the northern winter is the possibility of water pipes freezing. When they do, they crack and split because of the expansion of the ice inside. Exactly the same thing happens in nature. On icy mountains water soaks into pores and cracks in exposed rocks. When it freezes, the water turns to ice. The ice expands with a pressure that widens the pores and cracks, splitting the rocks apart and breaking up entire mountainsides. Masses of broken blocks that slope downwards from craggy peaks are a result of this destructive action. They are called talus slopes.

A similar mechanism brings stones to the surface of a garden in winter time. Water beneath a buried stone freezes more readily than that in the soil around because the stone absorbs its heat more quickly. Ice forming beneath the stone expands and pushes the stone upward. In permanently cold regions the whole soil surface is raised in a regular series of low humps. The stones brought to the surface collect in the troughs between the humps and produce a honeycomb pattern.

Ice can break and change rocks. It can also transport things. A glacier is a mass of ice that moves slowly under the influence of gravity. It is one of the most powerful means of natural transportation that there is. Much of the landscape of the Northern Hemisphere is formed from debris deposited by continent-wide glaciers during the ice ages of the last two million years.

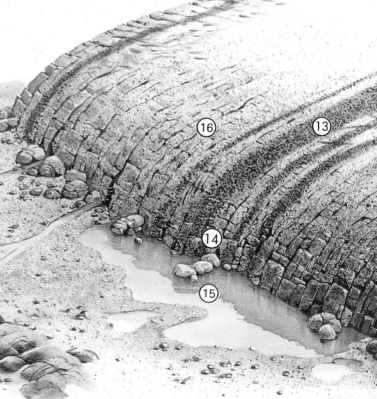

Features of a glacier

 1 Pyramidal peak
 2 Bergschrund, a crevasse
 3 Cirque hollowed out by glacier's weight
 4 Firn, a mass of compacted snow
 5 Surface crevasses
 6 Avalanche
 7 Seracs, columns of ice
 8 Deep crevasses
 9 Arête, a narrow ridge left between glaciers
10 Marginal crevasses
11 Pressure ridges
12 Lateral moraine, debris carried along edge
13 Medial moraine, formed by two lateral moraines
14 Snout
15 Meltwater
16 Ablation moraine, left as ice has melted
17 Glacier table, rock on a pedestal of ice
18 Englacial moraine, carried in the glacier
19 Subglacial moraine, under the glacier
20 Ice cave

42

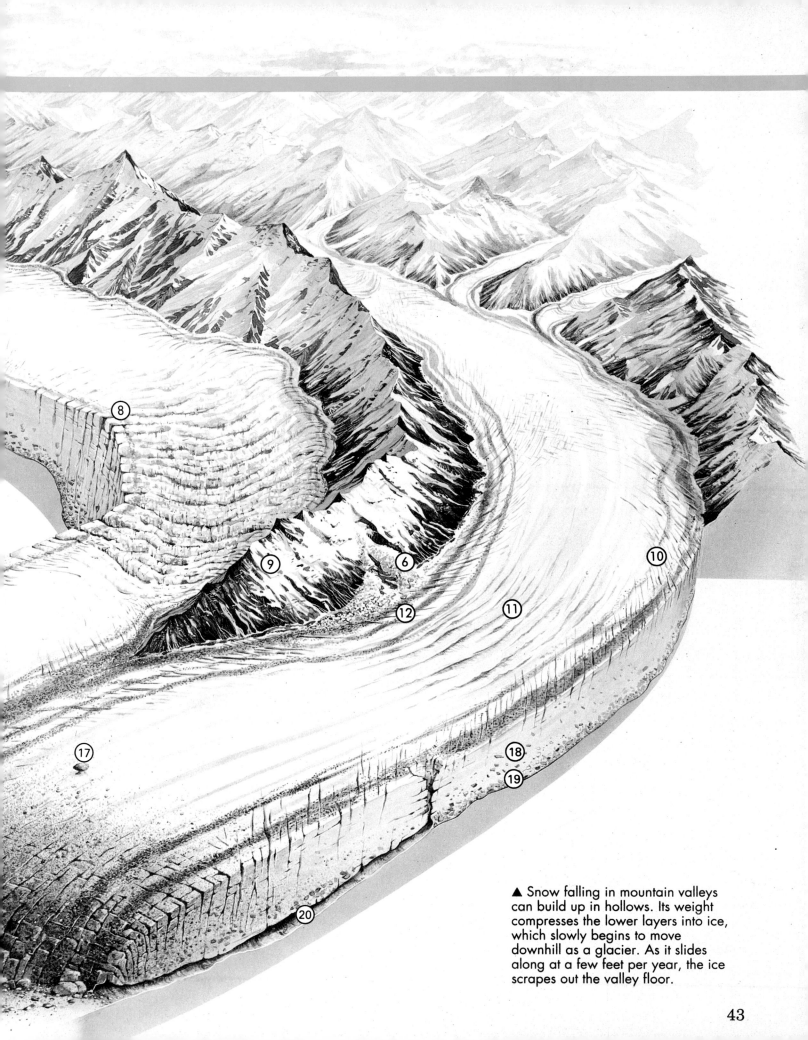

▲ Snow falling in mountain valleys can build up in hollows. Its weight compresses the lower layers into ice, which slowly begins to move downhill as a glacier. As it slides along at a few feet per year, the ice scrapes out the valley floor.

The changing scene

The landscape is never static. It changes from day to day, year to year, millennium to millennium. Most of these changes are natural. But with the coming of civilization, many have been caused by the activity of people.

Almost everything that civilization does has an impact on the surface of the Earth. About 5,000 years ago, irrigation was first practiced in the Middle East. Rivers were diverted to make dry desert areas fertile. The first cities were built, often constructed on artificial hills, for defense. The people of low-lying lands next to shallow seas often extend their farmland by walling off areas of sea and draining them. The Dutch have been doing this since the tenth century. Since the Industrial Revolution of the nineteenth century, vast areas of the landscape have been dug up for raw materials.

More spectacular are the unintentional changes in the landscape caused by human activities. Building breakwaters for harbors, or removing beach gravel for building, can alter sea currents. As a result, local seafronts can be washed away. Bad farming practices can alter the structure of the soil so that it falls apart and is washed away or blown away in the wind. In this way, fertile farmland can turn to desert very quickly. With the population increasing every year, the impact that civilization has on the environment becomes greater and more significant all the time.

▼ The world's biggest man-made hole in the ground is the Bingham Canyon Copper Mine in Utah. It has a depth of 774 m (2,540 ft.) and covers an area of 7.21 sq km (2.8 sq. mi.).

44

▲ In deserts, sand is blown along close to the ground at a height of about 1 m (1 yd.). It hits the base of rocks, eroding them into mushroom shapes.

▶ The sea is constantly eating away at the coastline. It undercuts cliffs and erodes headlands into isolated pillars of rock called stacks.

Igneous landforms

The shape of the landscape depends mainly on the type of rock it is made from. When an igneous rock fills a large crack, the structure is called a dike. As the surrounding rocks are worn away, the dike juts up like a wall across the scenery. An igneous rock forming a layer between beds of a sedimentary rock forms a structure called a sill. When this is eroded, it may look like a thick, hard sedimentary bed.

Sometimes the magma rising up a crack stops at a particular level and domes up the sedimentary rocks above it. This produces a laccolith, sometimes seen on the surface as a ringlike structure in the uplifted sedimentary rocks. When an ancient volcano is worn away, the solid igneous material in the vent may remain standing as a pinnacle (below), showing where the volcano once stood.

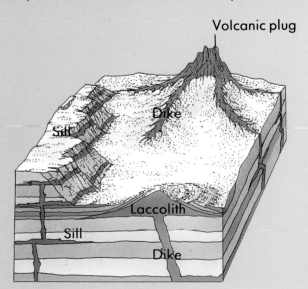

Volcanic plug
Dike
Sill
Laccolith
Sill
Dike

45

Part Two

Air and oceans

Earth is the only planet in the Solar System that supports life. The key to life on Earth is water. All living things, plants and animals alike, require water for their body structures and for making or digesting food.

Water exists as a liquid in vast quantities in the oceans, rivers, and lakes. It occurs as solid ice at the North and South poles. It is also present in small amounts as water vapor in the atmosphere. Water moves between the surface and the atmosphere in a never-ending cycle.

The shifting around the world of air masses containing more or less moisture is one of the major factors that determine the weather. The average weather pattern, or climate, at a particular place depends on many other factors as well. One of the most important of these is the temperature.

◄ Water is found in lakes, rivers, and oceans. Snow and ice are forms of frozen water. Clouds are droplets of water; the atmosphere holds water vapor. All plants and animals need water to live.

The oceans

We call our planet the Earth. It would be more appropriate if we called it the Water, because more than two-thirds of its area is covered by the seas and oceans. This is clearly visible in satellite photographs and the views that astronauts have from space. The overall color is blue, caused both by the effect of the atmosphere and by the vast areas of water beneath. It is the presence of all this water that has made life possible. All life processes involve water, and life evolved in the sea about 3.5 billion years ago. It was only about 500 million years ago that life moved out of the oceans and on to the land.

SPOT FACTS

• The oceans have a total area of 361,300,000 sq km (139,500,000 sq. mi.). This represents 71 percent of the total surface area of the Earth.

• The total volume of the oceans is 1,349,929,000 cu km (324 million cu. mi.).

• The average depth of the oceans is 5 km (3 mi.), but the deepest point is in the Marianas Trench east of the Philippines, at 11 km (7 mi.) deep.

• The temperature of surface ocean water ranges from 26°C (79°F) in the Tropics to -1.4°C (29.5°F), the freezing point of seawater, near the Poles. From a depth of 100 m (330 ft.) to approximately 1,000 m (3,300 ft.), ocean temperature is about 5°C (41°F). Below this depth it drops slowly to almost freezing.

• If removed from the sea, the dissolved salt would cover the land to a depth of over 150 m (490 ft.).

• The sea is so huge that several valuable metals are abundant, but it is difficult to extract them. The oceans are estimated, for example, to contain 10 billion metric tons of gold; yet the concentration is so low that it is impossible to recover any of it.

• It is estimated that 90 percent of all photosynthesis and release of free oxygen takes place in the oceans.

• While speeds of surface currents can be as much as 250 cm (98 in.) per second, speeds of deep currents vary from just 2 to 10 cm (about 1 to 4 in.) per second or slower.

• Current annual food harvests from the oceans amount to about 52.5 million metric tons of fish, 1.5 million metric tons of large whales, and about 700,000 metric tons of seaweed.

• Currently, the major minerals being obtained from seawater are magnesium, bromine, and sodium chloride (common salt).

• The ocean floor gives us sand and gravel for the building industry. Small quantities of diamonds are found in some submarine gravel bars.

• Much attention has recently focused on manganese nodules, which are round concretions on the deep ocean floor. These nodules contain about 20 percent manganese, 10 percent iron, 0.3 percent copper, 0.3 percent nickel, and 0.3 percent cobalt. These are all valuable minerals that have not yet been obtained to any great extent from the seafloor.

• Offshore oil and gas wells currently supply about 17 percent of global petroleum production. Most of these wells are in the shallow waters of the continental shelves. In the future, deep-sea drilling technology is expected to exploit petroleum reserves on the outer continental margins.

• The pollution of the oceans by oil and chemical spillage and sewage disposal has focused world attention on the need for controlled use of resources and sensible disposal of waste.

• Other pollution worries are the effects of pesticides on marine life, increasing lead levels in surface waters, and the pumping of waste hot water from power plants into the sea with undesirable effects.

The water

The water in the sea tastes very salty. Drinking it is likely to make you nauseous. Seawater is not pure, but contains a large quantity of salts that have been dissolved from the rocks of the Earth's crust. The actual salinity, or saltiness, varies from place to place. It is highest in warm enclosed seas, where water is constantly being evaporated. Seawater is least salty in the cold northern and southern oceans, where it is diluted by rain and melting ice. It is also less salty where great rivers like the Amazon or Niger flow into the ocean. However, the proportions of the different salts present are constant throughout the world.

The temperature of seawater varies a great deal over the surface of the ocean, but the temperature beneath remains a fairly constant 4–5°C (15–19°F). The slightest variation in temperature can trigger ocean currents.

Combinations of salinity, temperature, currents, and other factors determine the amount of life present in any ocean region. Vigorous life is confined to the surface and a few hundred feet below it. There the Sun shines into the water, and plant life, some of it microscopic, can grow. Small invertebrates feed on the plants, and fish feed on these. A whole ecosystem is supported. In the dark depths the Sun has never shone. The only life consists of creatures that feed on dead organic debris, which rains down from the more fertile layers above, or that prey on one another.

▼ The distribution of land and sea on our globe is not even. Most of the ocean is on one side, in the Pacific. This is an accident of plate tectonics. Some 200 million years ago the inequality was even greater, with all the continents fused together and the rest of the world covered with a single ocean, called Panthalassa.

Seawater composition

Much of the gas belched out of volcanoes is water vapor. It is actually recycled seawater, which has been carried down into the depths of the crust by plate tectonic movement. The salinity of ocean water varies usually between 33 and 38 parts per thousand. The dissolved salts contain nearly all the chemical elements. The most abundant are sodium and chlorine, which together make up common salt.

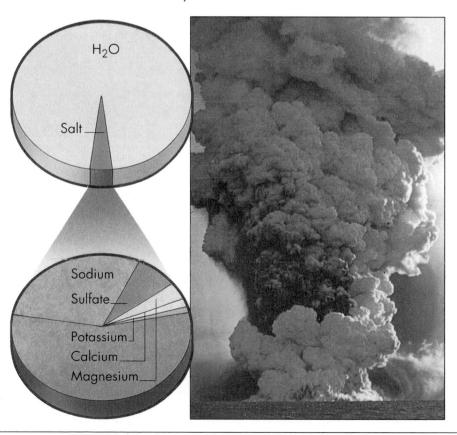

The ocean surface

The continents are large islands, mostly quite separate from each other. As a result, the oceans form a single continuous body of water. In some places, such as the Bering Strait, the gaps between the continents are narrow; in others, such as the East Indies, they are choked with islands. But nowhere is there an area of ocean that is completely isolated from any other. Political problems aside, we can sail from a port on one continent to any other seaport in the world. For convenience, however, we often talk about the seven seas. These are the North Atlantic, South Atlantic, Indian, North Pacific, South Pacific, Arctic, and Antarctic Oceans.

The surface of this vast waterway is constantly in motion, notably through the movements of ocean currents. Vertical currents are set up by convection. This happens because a cold mass of fluid is denser than a warmer mass of the same fluid and sinks through it. Cold water from melted ice at the North Pole sinks through the surrounding warmer water. It then travels for thousands of miles as an undersea current along the bed of the Atlantic. On the surface, however, it tends to be the prevailing winds that power the ocean currents. The Trade Winds are those that blow toward the Equator from the northeast and southeast. The surface waters blown by them produce an overall westward flow of equatorial water known as the equatorial currents.

When this water reaches a continent, it sweeps north and south, producing warm currents along the east coast of that continent. The Gulf Stream in the North Atlantic, the Kuroshio in the North Pacific, and the Australian Current in the South Pacific are examples. The movement is completed when the water sweeps toward the Equator along the eastern edge of the ocean and joins up with the beginnings of the equatorial currents. These include the California Current and the Peru Current in the eastern Pacific Ocean.

Hence, the world pattern of ocean currents is based on a vast system of circular movements, or gyres, each occupying half an ocean. The movement of warm currents along cold continental edges, and cold currents along warm continental edges, helps to modify the climates in these coastal areas.

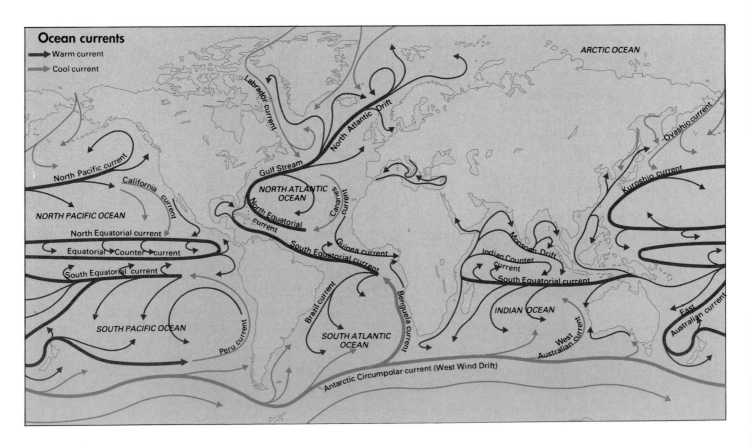

50

▲ Penguins and flat-topped icebergs typify the cold Antarctic Ocean. The icebergs are made from glaciers that have broken off at the edge of Antarctica. The ice of the North Pole, on the other hand, is formed on the sea's surface.

◄ Prevailing winds drive the ocean currents. The Westerlies blowing in the far south produce the cold circumpolar current called the West Wind Drift. This separates the Antarctic from warmer waters farther north and keeps Antarctica frozen. Elsewhere, the ocean gyres bring warm or cool currents to the edges of the continents.

▶ During the ice ages of the last 2 million years the sea level changed. So much water was locked up in the expanded ice cap at the North Pole that the volume of liquid water left was less than it is now and the sea level was lower. Land appeared where there is now shallow sea. At the same time in the far north, the great weight of the ice on the land pushed the land downward, and sea levels there were higher than they are today. The ice came and went several times, and we can often see the different sea levels as "raised beaches." These are banks a few feet above sea level that mark the ancient shorelines.

Today

An ice age

51

The ocean floor

Under the sea is a varied landscape that is rarely seen by human beings. It is a landscape of mountains, volcanoes, and broad, flat plains. Only since the 1960s have we really begun to understand it.

There are a number of different zones of the ocean floor. At the edge is the continental shelf. This is merely the water-covered edge of the continent. If we drill through the sediment on the continental shelf, we find continental crust beneath it, not oceanic crust. The continental shelf can be hundreds of miles broad where it is at the edge of an old continent, as for example in the North Sea and Hudson Bay. But the shelf is very narrow or nonexistent where new mountain ranges are being pushed up along a coastline, as along the western coast of South America.

At the edge of the continental shelf the seabed slopes downward into the depths. This feature is known as the continental slope, and it marks the edge of the continent itself. In some places – usually offshore from the mouths of large rivers – the continental slope is cut by vast canyons, wider and deeper than any canyon on land. These have been eroded out by debris swept down from the rivers.

At the foot of the continental slope lies the continental rise. This is not as steep as the

▼ Coral reefs are found only in shallow tropical waters. Coral builds out from an island (inset) in a shelf called a fringing reef. Over thousands of years, the island may sink as the coral builds up. A lagoon then separates the reef, now a barrier reef, from the dwindling island. When the island sinks completely, it forms an atoll, which is a ring of coral surrounding a lagoon.

continental slope and is built up from debris brought down from the continental shelf above. Vast fan-shaped masses of sediment may form at the mouths of the submarine canyons. Perhaps the largest are from the outflowings of the Ganges and Indus Rivers.

The abyssal plain is the floor of the ocean itself. With true oceanic crust as its foundation, it extends between the continents. It is covered with a fine muddy deposit called ooze. This is made up of the remains of millions of sea creatures that have settled on the dark bottom over many millions of years. There is no debris from the land here. Across the abyssal plain rise the volcanic ocean ridges. Occasionally the plain drops away into the ocean trenches, the deepest points of the ocean, where the old crust is continually being destroyed.

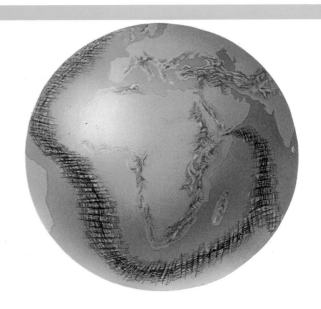

▲ The mid-ocean ridges are the most extensive geographical feature on the Earth. They run through the beds of all the oceans without a break.

Geography of the ocean floor

The continental shelf is made of continental crust. It reaches out from the shore to a depth of about 150 m (500 ft.). At its outer edge it forms the continental slope, with a gradient of between 3 and 20 degrees. The slope, and the continental rise at the bottom, gives way to the abyssal plain at a depth of about 4,000 m (13,000 ft.). The mid-ocean ridge rises from the abyssal plain to heights of 500–1,000 m (1,640–3,280 ft.).

Volcanoes erupt in the rift valley that exists along its crest. The ridge volcanoes may occasionally reach the ocean surface, but more often they are completely submerged. Seamounts on the abyssal plain, some with flat eroded tops, are the remains of these volcanoes. The ocean trenches, in which one part of the Earth's crust is being swallowed up beneath another, are the deepest points on Earth.

▼ The volcanic activity in an ocean ridge is dramatically shown by the presence of "smokers." These are jets of hot water that burst from the Earth's crust at great depths. The "smoke" is caused by fine particles of minerals in the water.

The ocean floor

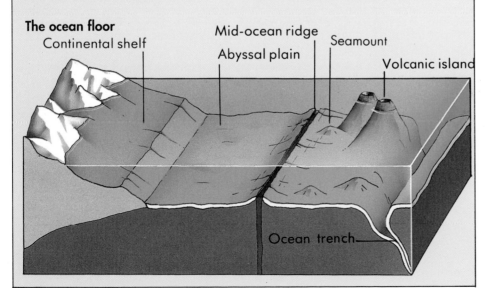

Continental shelf • Mid-ocean ridge • Abyssal plain • Seamount • Volcanic island • Ocean trench

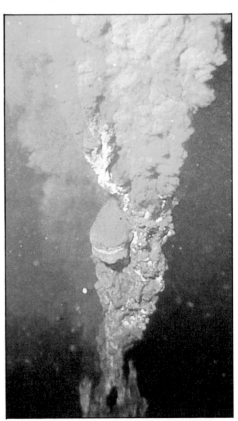

53

The water cycle

More than two-thirds of the Earth's surface is covered with water. By far the greatest volume of this – 97.2 percent – is contained in the oceans. The remainder is found as water vapor, fresh water, and ice. Water occurs as vapor in the atmosphere, as ice in the glaciers and ice caps, and as liquid water flowing in rivers, standing in lakes and swamps, and absorbed into rocks and soil as groundwater. Water vapor in the atmosphere condenses to form raindrops, which fall to the ground. Rainwater collects to form streams and rivers, which flow to the sea. If it were not for water, which is available in most places on the Earth, life would be impossible on our planet.

SPOT FACTS

• The Earth contains 1.4 quintillion ("14" followed by 17 zeroes) metric tons of water.

• Only one millionth of the Earth's water is found flowing in rivers. A hundred times more than that is held in lakes.

• About 80 percent of all the lake water on Earth is held in just 20 lakes.

• Water is the only substance that occurs at ordinary temperatures in all three states of matter, that is, as a solid, a liquid, and a gas.

• There are 20 million sq km (8 million sq. mi.) of ice in the form of ice caps and glaciers.

• River water is not pure, but it is not as salty as seawater. Its salinity is measured in parts per million rather than parts per thousand. The main dissolved substance in river water is calcium carbonate.

• The longest river in the world is the Nile, at 6,680 km (4,150 mi.).

• Because of its capacity to dissolve numerous substances in large amounts, pure water rarely occurs in nature.

• Raindrops usually have a diameter greater than 0.5 mm (0.02 in.). They range in size up to about 3 mm (about 0.13 in.) in diameter.

• Rain falls at up to 7.6 m (25 ft.) per second, depending on the size of the drops.

• The world's heaviest average rainfall, about 11,873 mm (about 467 in.) per year, occurs at Mawsynram, in India. The average annual rainfall around Antofagasta and Arica in Chile is less than 0.1 mm (0.004 in.).

• In the United States, the southeast of the country has the heaviest average rainfall: 178 cm (70 in.).

• In 1970, 38.1 mm (1.5 in.) of rain fell in one minute at Barst, Guadeloupe, in the Caribbean.

• Totally pure water does not freeze below 0°C (32°F) unless disturbed with a small ice crystal.

• The first dam on record was built in Egypt in about 4000 BC to divert the Nile and provide a site for the city of Memphis.

• The Nurek Dam in Tajikistan, with a height of 300 m (984 ft.), is the highest dam in the world.

• Cloud particles vary in diameter from 5 to 75 microns (0.0005 to 0.008 cm/0.0002 to 0.003 in.).

• The scientific study of clouds started in 1803, when a method of cloud classification was developed by the British meteorologist Luke Howard (1772–1864).

• One of a strange but beautiful group of clouds is the nacreous, or mother-of-pearl, cloud, found 19 to 29 km (12 to 18 mi.) high. Another is the noctilucent cloud, found 51 to 56 km (32 to 35 mi.) high. Both these very thin clouds can be seen only between sunset and sunrise, and are visible only in high latitudes.

Moving water

Physical conditions on the Earth's surface ensure that water can exist in all three of its possible states. It can exist as a gas (as water vapor); it can exist as a liquid (in the form that we most often see it); or it can exist as a solid (in the form of ice). It takes a relatively small shift of conditions such as temperature or pressure to change water from one state to the next. One cold night can turn a liquid pond into solid ice, or one hot day can evaporate a large puddle to vapor, leaving the surface dry.

Water is continually evaporating into the atmosphere from open bodies of water, such as seas and lakes. Then it condenses and falls as rain. The rain that falls on the land runs off the surface or sinks into the rocks and soil to form the groundwater. It reemerges at springs and forms the beginnings of streams and rivers, and flows back to the lakes and oceans. This whole process is known as the water cycle.

The cycle has many side branches as well. Water that falls on the ground may quickly evaporate. Groundwater may be drawn up through plants and evaporated from leaves.

The water cycle

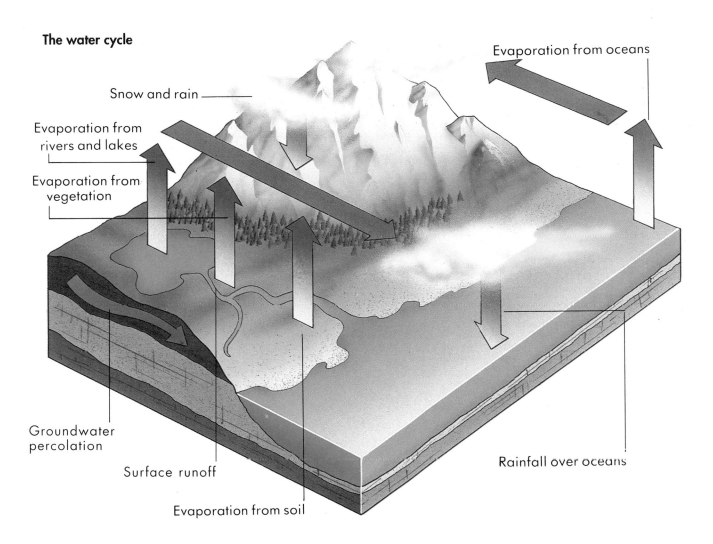

▲ Water goes round and round in the water cycle. The main upward movement is by evaporation. Water evaporated as vapor from oceans, rivers, and lakes passes into the atmosphere. Some water vapor is also "breathed out" by plants. High in the air, water vapor condenses to form clouds. Sideways movement occurs when clouds are blown along by the wind. Finally, to complete the cycle, downward movement takes place when it rains. When it is cold, rain falls as snowflakes or hailstones, and lies as snow and ice. Sometimes snow and ice can even change directly into vapor without passing through a liquid phase.

55

Water resources

We all need water to keep us alive. Each of us needs at least 2 liters (about 2 qt.) every day just to keep the body working. In Westernized societies the daily consumption is much greater because of the water used for washing and the demands of industry and agriculture. On average it takes 2,000 liters (2,110 qt.) of water a day to support one person in an industrial nation. It is no wonder that water is one of the most valuable of the Earth's resources.

The surface of the Earth is largely covered by water, but unfortunately most of this is unusable or in the wrong place. Many rapidly expanding population centers lie in areas where there is very little water because the climate is too dry and the rainfall irregular. Other expanding centers of population lie where the presence of water is a source of problems, such as by tropical rivers that are likely to flood and promote the spread of disease.

Much of the usable water is obtained from rivers. The irregular or seasonal flow of a river can be regulated and controlled by using dams.

Built across the flow of a river, a dam traps the water and fills the valley behind it to produce a reservoir. The larger a dam becomes, however, more and more problems are associated with it. A heavy structure on unstable ground could collapse, and with several million metric tons of water behind it the result would be disastrous. Reservoirs may also silt up, because the original flow of the river is disrupted. The floor of the reservoir builds up so that it becomes shallow and can hold only a small fraction of the original water. The stagnant surface of the water may also promote the growth of waterweeds that both choke waterways and make the water unmanageable.

The other great water source is groundwater,

▶ The Kariba Dam, at the border between Zambia and Zimbabwe, in southern Africa. The dam creates the large reservoir of water called Lake Kariba.

▼ The plain of the Indus River in India is irrigated by a complex system of channels that bring the spring meltwater from Himalayan glaciers to the fields.

the water that has soaked into the rocks and soils of an area. This can be extracted by digging wells or boreholes and installing pumps. In the driest areas, however, this does not represent an inexhaustible supply. Much of the water that is being pumped up in northern Africa is actually rain that fell in the Ice Age. Once that has gone, there will be no more.

As the world population grows there will be bigger and more elaborate schemes for distributing water to where it is needed. Each will bring its own problems.

▶ The constant cycle of water is particularly apparent in the daily downpours that occur in the areas of tropical forest that straddle the Equator.

Groundwater

Oases

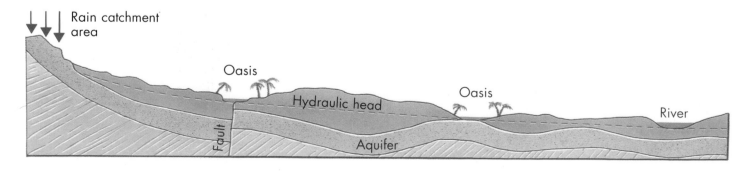

Rain catchment area

Oasis

Fault

Hydraulic head

Aquifer

Oasis

River

When rain falls on the soil, some of it washes off as surface flow. The remainder sinks into the soil and becomes groundwater.

Soils tend to be quite loose and have large air spaces and pockets between the particles. These are easily filled up with water, in a similar way to the holes in a sponge. The rocks beneath are more compact, but usually contain air pores as well. If the pores are connected together, water can soak through. We say that the rock is permeable. At some distance below the surface the rocks are so compact that they have no spaces in them and water cannot penetrate. The rock is impermeable. When the rock above this level is full of water, it is said to be saturated. Above the saturated zone is a region in which the water is seeping downward. This is the zone of intermittent saturation.

The upper boundary of the saturated zone is called the water table. Its level varies from time to time, being higher in wet weather and very low in times of drought. When wells are drilled, they are driven down to below the water table. Water from the surrounding rocks gathers in the bottom of the hole. Where the water table reaches the surface, the water leaks out and becomes a spring. Other springs form when the water from the saturated zone seeps upward through cracks or faults.

The movement of water through underground rocks is slow, usually taking years. Once underground, the water is protected from evaporation, and its passage through the pores of the rocks filters it so that it is usually quite clean. Accordingly, underground water is a valuable resource, although there may be some salts and minerals dissolved in it.

It is estimated that about 62.5 percent of the world's fresh water is present in the form of groundwater in the rocks.

◀ Groundwater is widely exploited in areas where there is little rainfall and few rivers. In desert oases, such as this one at Taghit in Algeria, the water table reaches the surface and a freshwater lake forms. The water is used for drinking and irrigation.

▶ In most instances the water table lies well below the surface, and the water has to be brought up by artificial means. Here oxen are used to turn a wooden gear that brings water to the surface by a bucket wheel for distribution to the groves of date palms. Desert peoples have become skillful at drawing water and transporting it to the fields by irrigation systems.

Artesian well

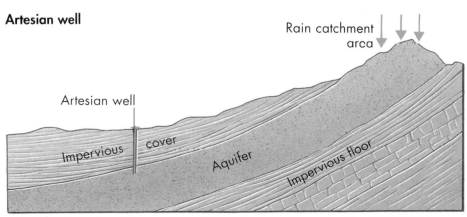

Rain catchment area

Artesian well

Impervious cover

Aquifer

Impervious floor

◀ A rock that contains groundwater is called an aquifer. Different kinds of wells can be drilled to reach the water in an aquifer. One type is the artesian well. Exposed rocks in a hilly area absorb rainwater, which seeps into the rock. In the nearby lowlands, a well is drilled through the overlying impermeable bed into the aquifer to reach the water. The pressure of the water coming down the slope of the aquifer pushes the water up the borehole until is gushes out at the surface.

Clouds and rain

The water vapor in the atmosphere makes up only about 0.001 percent of the total water supply of the world. But without it all life would be confined to the sea. The proportions of most of the gases in the atmosphere are constant wherever we go, but the proportion of water vapor varies.

The concentration of vapor in the atmosphere depends largely on the temperature and pressure. When the atmosphere is holding as much vapor as it can at a particular temperature and pressure, it is said to be saturated. Any decrease in temperature or pressure will turn some of the vapor back to liquid. Because the atmosphere is always in motion, the physical conditions are changing all the time. So is the amount of water in the air.

Such changes can be caused by the wind blowing from a warm region to a cold region. Convection currents may lift warm air up to heights where the pressure is less. Wind blowing off the sea can carry moisture up a hillside, where the air becomes cooler as it rises. When these things happen, the vapor in the air turns to water in the form of tiny droplets. These are too light to fall to the ground and remain suspended as clouds.

Different types of clouds form under different conditions. At very great heights, above about 7,000 m (23,000 ft.), the conditions produce clouds of ice crystals rather than droplets. At other heights the water droplets form layer clouds, called stratus; or heaped clouds, called cumulus; or a combination of the two. Thunderclouds have such strong convection currents that the main cloud mass is made of water droplets, while its crown is of ice.

Cooling and decreasing the pressure still further cause the droplets to mass together into larger drops, and these fall as rain.

▼ A satellite view of a hurricane. The rapid heating of surface ocean waters in tropical regions causes hurricanes to form. Turbulent and constantly changing conditions are found in the hurricane. Winds spiral in toward a low-pressure area and thick banks of clouds form as the pressure becomes less and less. Torrential rains fall as the pressure reduces still further.

▶ Clouds form in the troposphere at heights of up to about 11 km (7 mi.). Stratus and stratocumulus are the lowest clouds, often forming gray sheets that bring light rain and fog. Cumulonimbus clouds can extend up to about 6 km (4 mi.). They form huge towers in the sky and bring thunderstorms with rain, snow, or hail. The highest of all are the thin, wispy cirrus clouds.

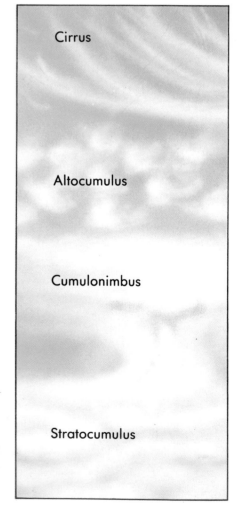

Cirrus

Altocumulus

Cumulonimbus

Stratocumulus

▲ Heavy seasonal rains called monsoons develop around the northern Indian Ocean. In the winter the Asian landmass cools and dry air flows seaward. In the summer the continent heats up and draws in wet air from the ocean. The moisture falls as rain over the land.

◄ Thunderstorms develop where a patch of air heats up quickly and rises. Raindrops form under the reduced pressure at high altitudes, and are carried up and down by the strong currents. The growing drops become too big to be stable and split into smaller drops, producing an electrical discharge. A strong charge builds up in the cloud and flashes to Earth as lightning.

Life-giving atmosphere

The Earth's atmosphere is an ocean of air that surrounds the planet. Air is a mixture of gases, mainly nitrogen and oxygen. We cannot see it, or taste it, or smell it, yet air is vital for life. Layers of gases high in the atmosphere shield us from harmful radiation from the Sun. But the increasing destruction of these protective layers caused by pollution is altering the Earth's climate. We take oxygen from air into our bodies with every breath, and without it we suffocate. Oxygen in the air is also needed for fuels to burn. And air does have substance. Without it, birds and aircraft could not fly.

SPOT FACTS

• The atmosphere weighs about 5 quadrillion ("5" followed by 15 zeroes) metric tons.

• At sea level, the weight of the atmosphere exerts a pressure of 1.05 kg per sq cm (about 15 lb. per sq. in.).

• We can study changes in climate over the last half million years by looking at the composition of the ancient ice in Antarctica and Greenland.

• The greenhouse effect may help to avert another ice age.

• The density of dry air at sea level is about 1/800th the density of water.

• Among natural air pollutants, only the radioactive gas radon is considered a health threat. Radon seeps into the basements of homes built on rocks containing uranium. According to recent estimates by the United States government, 20 percent of U.S. homes harbor radon concentrations that are high enough to pose a risk of lung cancer.

• All living things need oxygen. It reacts with food molecules, such as fats and carbohydrates, to release energy, along with carbon dioxide and water. This process is called respiration, and occurs within the cells of plants and animals.

• The oxygen in the air is continually replenished by green plants. In daylight hours they take in carbon dioxide from the air, combining it with water to produce sugars. Oxygen is the byproduct.

• Air can be separated into its constituent gases by liquifying it and then by fractional distillation.

• Argon, which forms one percent of the air, is used in large quantities in the manufacture of electric light bulbs. Bulbs are filled with argon to prevent the filaments from burning up, which they would do if the bulb contained air.

• The problem of acid rain began during the Industrial Revolution, and it has been worsening ever since. All industrialized nations are now committed to reducing sulfur dioxide emissions from power plants. The fitting of catalytic converters to automobiles is another weapon against the problem.

• An altimeter is a device that uses reducing air pressure with height to determine altitude. Modern aircraft tend to use radio altimeters that measure the time taken for a radio wave to bounce off the ground.

• When a space vehicle reenters the atmosphere, air friction produces very high temperatures. The Space Shuttle is protected from burning up by special ceramic tiles.

• The troposphere and most of the stratosphere have been explored by means of balloons equipped with instruments to measure the pressure and temperature of the air. A radio transmitter sends the data to a receiving station on the ground.

• The record drop for a parachute descent was set in August 1960, when Captain Joseph Kittinger bailed out of a polyethylene plastic balloon at 31,330 m (102,800 ft.) over Tularosa, New Mexico.

The air we breathe

The atmosphere is a thin layer of air that surrounds our planet. On a small model of the Earth its thickness would be hardly noticeable – about as thick as the skin of an apple. But its weight at ground level gives us the air pressure under which we and all other land-dwelling creatures evolved. The higher up we go, the thinner the atmosphere becomes. At a height of several hundred miles it fades away into the vacuum of space.

The atmosphere is a mixture of gases, mainly nitrogen and oxygen, with about four parts of nitrogen to every one of oxygen. The remainder – just over one percent of the whole – consists of argon, with tiny amounts of carbon dioxide and rare gases like helium and neon. These rare gases, together with argon, are called noble or inert gases as they do not take part in chemical reactions. The proportions of all these gases tend to remain the same all the time.

▶ Most of the atmosphere consists of nitrogen, but about 21 percent of it is oxygen generated by the action of plant life. The oxygen makes Earth's atmosphere different from those of other planets.

▼ Mountaineers who climb to high altitudes generally carry a supply of oxygen to breathe. At high altitudes the air pressure is less, and so there is less oxygen available in the air.

There are, however, a few variable components in the atmosphere. The most important variable is water vapor. It can be almost absent in desert areas but can reach a concentration of about three percent in very humid regions. The presence of water vapor is essential to life. Sulfur dioxide is another variable. This is not essential to life, and can in fact be quite harmful. It is produced in large amounts by volcanoes and by burning fossil fuels such as coal and oil.

At a height of between 15 and 50 km (9 and 30 mi.) there is the so-called ozone layer. Ozone is a type of oxygen. High-energy ultraviolet radiation from the Sun affects oxygen in the layer and turns it into ozone. This reaction absorbs ultraviolet radiation and prevents most of it from reaching the Earth's surface. The layer is important because too much ultraviolet radiation would be harmful to living things.

Composition of air

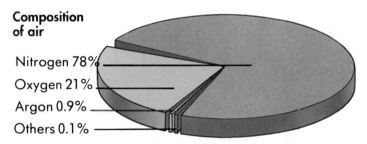

Nitrogen 78%
Oxygen 21%
Argon 0.9%
Others 0.1%

63

The atmosphere's structure

The layers of air that form the atmosphere stretch upward above our heads for about 700 km (over 400 mi.). At that height the air is extremely thin. At even higher altitudes the atmosphere fades away into the vacuum of space.

The "thickness" of the atmosphere is called its density. Air is densest near the surface of the Earth. It is less dense at the tops of tall mountains. Atmospheric pressure is also greatest near the ground. The pressure is caused by the weight of layers of air pressing down from above.

The atmosphere can be divided into a number of layers each with its own properties. The lowermost 11 km (7 mi.) or so is called the troposphere. Although it is quite a thin layer, the troposphere contains, under pressure, the greatest proportion of air by mass. All the physical activities that affect the weather take place in this region.

The top of the troposphere is a theoretical boundary called the tropopause. Above this lies the stratosphere, extending up to about 50 km (30 mi.). Most military and long-distance aircraft operate in this region. The ozone layer, within this region, absorbs much of the Sun's energy. As a result, the temperature is quite high in the upper stratosphere.

Above the stratopause – the upper limit of the stratosphere – stretches the mesosphere, up to about 80 km (50 mi.). The temperature there is low and the air is thin. But it is still thick enough for meteorites to burn up as they pass through it. Beyond its upper boundary, the mesopause, the mesosphere gives way to the thermosphere. This is another region of relatively high pressures caused by absorption of solar radiation. Then comes the exosphere, which eventually fades away to nothing over 700 km (435 mi.) above the surface of the Earth.

▶ Layers of the atmosphere. The weight of the air pressing down on itself compresses the lower layers. As a result the lowermost layer, the troposphere, contains 80 percent of the atmosphere by mass. But it occupies a volume of only 1.5 percent. Above the stratosphere there is only one percent of the mass of air, but this is spread through 93 percent of the volume. The two diagrams (right) compare the composition of the atmosphere in terms of mass and in terms of volume.

▼ Radiation from the Sun includes light, heat rays, and ultraviolet radiation. Over 30 percent of the radiation reaching the Earth is reflected back by the atmosphere, by clouds, and by the ground. Most of the remaining 70 percent is absorbed. The ground reflects the least radiation and absorbs the most.

SOLAR RADIATION 100%

21% reflected by cloud

6% reflected by atmosphere

5% reflected by ground

15% absorbed by atmosphere

3% absorbed by cloud

50% absorbed by ground

Composition by mass

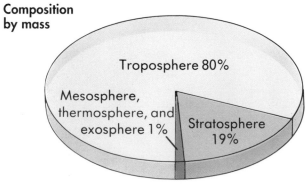

Troposphere 80%

Mesosphere, thermosphere, and exosphere 1%

Stratosphere 19%

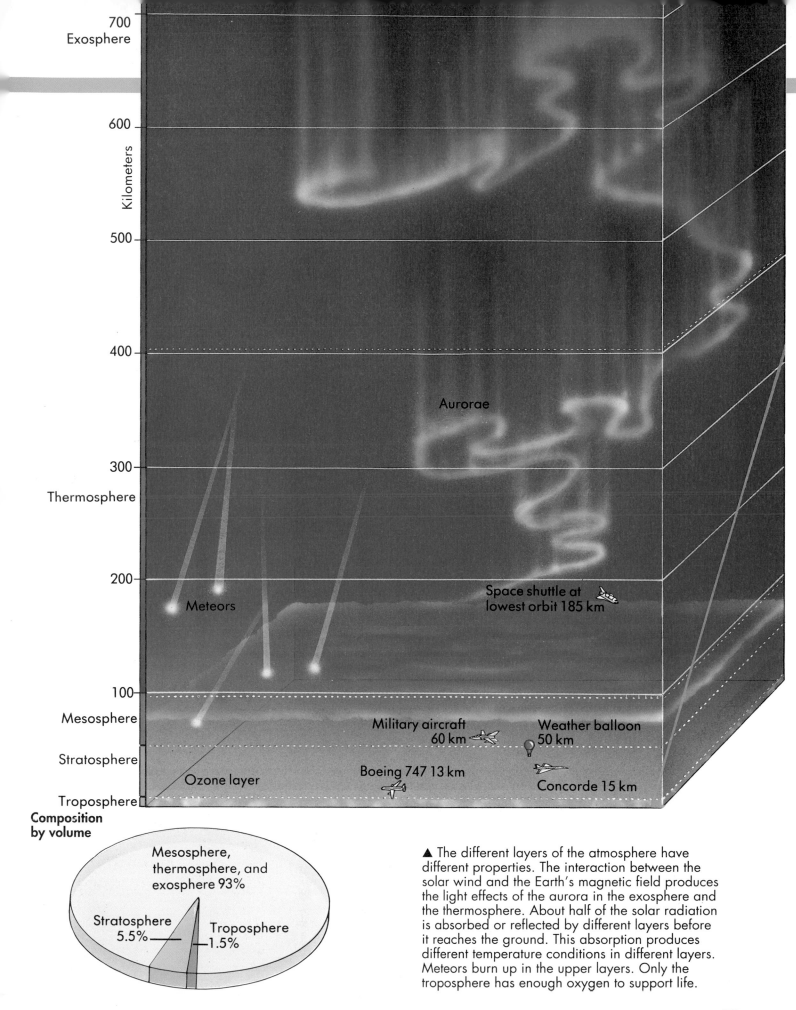

700
Exosphere

600

Kilometers

500

400

Aurorae

300

Thermosphere

200

Meteors

Space shuttle at
lowest orbit 185 km

100

Mesosphere

Military aircraft
60 km

Weather balloon
50 km

Stratosphere

Boeing 747 13 km

Concorde 15 km

Ozone layer

Troposphere

**Composition
by volume**

Mesosphere,
thermosphere, and
exosphere 93%

Stratosphere
5.5%

Troposphere
1.5%

▲ The different layers of the atmosphere have different properties. The interaction between the solar wind and the Earth's magnetic field produces the light effects of the aurora in the exosphere and the thermosphere. About half of the solar radiation is absorbed or reflected by different layers before it reaches the ground. This absorption produces different temperature conditions in different layers. Meteors burn up in the upper layers. Only the troposphere has enough oxygen to support life.

65

The changing atmosphere

The atmosphere that formed as the Earth first solidified was very different from the atmosphere of today. Even now it gradually continues to change.

The first atmosphere probably consisted mostly of carbon dioxide, nitrogen, hydrogen, carbon monoxide, and inert gases. The strong solar wind would immediately have blasted much of this away into space. Then as the Earth began to solidify, gases were emitted from the cooling rocks and built up the next atmosphere. These gases consisted largely of carbon dioxide, with some nitrogen, hydrogen, and traces of argon. Volcanoes continued to bring up water vapor, carbon dioxide, hydrogen sulfide and nitrogen. The Sun's energy broke down some of the water vapor into hydrogen and oxygen.

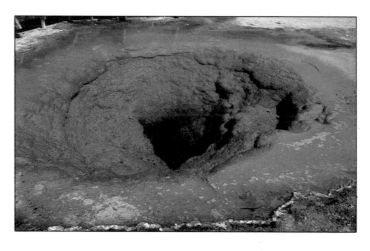

▲ This colony of green algae is living on chemicals in the hot, oxygen-free water of a hot spring. The first living things must also have lived in an oxygen-free environment. Their biochemistry was what scientists call "anaerobic."

It also converted some of the oxygen into ozone to produce an ozone layer early in Earth's history. The first oceans formed when so much water vapor was released that the atmosphere could not hold it all. The vapor condensed into clouds, and it began to rain.

The next major change in the atmosphere took place when carbon dioxide dissolved in the early oceans. We can tell that this was happening because we can find early rocks that contain calcite. This mineral was formed from carbon dioxide dissolved in seawater. The level of carbon dioxide in the atmosphere fell from about 80 percent to its present level of about 0.05 percent by about one billion years ago. Meanwhile, the hydrogen in the atmosphere was leaking off into space. It was too light to be held firmly by the Earth's gravity. As a result of the loss of these gases, the proportion of nitrogen gradually grew until it reached its present proportion of about 80 percent.

Enter oxygen

The most important change to the atmosphere began about 2.5 billion years ago. Before this time there was very little oxygen in the atmosphere. We know this because of the rocks that formed at the time. The iron in them formed minerals that were poor in oxygen. But if there had been free oxygen in the atmosphere, the iron would have formed rust-red minerals. The

► Colonies of anaerobic bacteria. Many types of these live in stagnant lakes and estuaries. The colonies shown here are living in black mud that contains no oxygen.

66

▲ The burning of coal, oil, and wood has an effect on the composition of the atmosphere. Burning uses up oxygen from the air and releases carbon dioxide. Water vapor is also produced by burning these fuels. Carbon dioxide and other gases together give rise to what is known as the greenhouse effect. The Sun's rays pass through the atmosphere and warm the Earth in the normal way, but the excess heat cannot escape back out again. Sulfur dioxide is also produced by burning coal, and this reacts with the moisture in the air to produce sulfuric acid. The result is acid rain, which damages plants and poisons lakes wherever it falls.

oldest "red beds," with iron oxides, date from about 2.5 billion years ago. At about this time primitive living things in the sea were beginning to use the energy of the Sun to make their food. They were the Earth's first plants. A by-product of this energy-changing activity was the generation of oxygen. The oxygen gradually built up until it reached its present level about 500 million years ago.

Now the atmosphere is changing again. Large-scale forest clearance cuts down the amount of oxygen produced, industry adds carbon dioxide to the atmosphere, and many processes disrupt the ozone layer. The long-term effects of these changes on climate and on living things have yet to be seen.

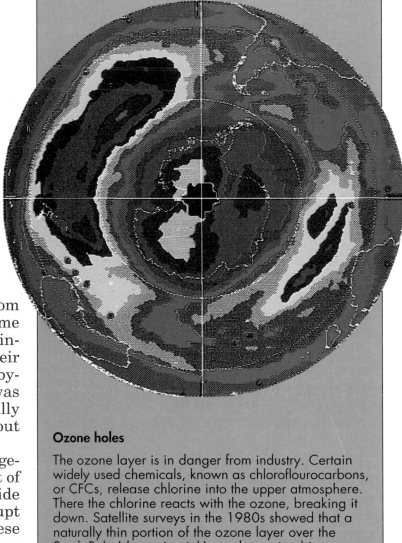

Ozone holes

The ozone layer is in danger from industry. Certain widely used chemicals, known as chloroflourocarbons, or CFCs, release chlorine into the upper atmosphere. There the chlorine reacts with the ozone, breaking it down. Satellite surveys in the 1980s showed that a naturally thin portion of the ozone layer over the South Pole (shown in pink) was becoming bigger.

Weather and climate

There is a difference between weather and climate. In a particular geographical region, climate is the overall result of the atmospheric conditions, averaged over a long period of time. The conditions include temperature, atmospheric pressure, and wind patterns. Weather, on the other hand, depends on the day-to-day variations of these conditions. Weather is important to agriculture, fishing, transportation, and many other human activities. This is why meteorology – the scientific study of weather – is a very important science. Because so many people depend on them, weather reports have to be accurate.

SPOT FACTS

• Frozen rain can be carried up and down in a thundercloud until it forms large hailstones as big as tennis balls.

• Jet streams – narrow belts of high winds that circle the globe at heights of about 10 to 20 km (about 6 to 12 mi.) – can reach speeds as high as 550 km/h (about 340 mph).

• When the Vikings discovered Greenland, they named it for the green plants that thrived there. A much warmer climate than today's prevailed at the time.

• Nuclear war could lead to a "nuclear winter," during which sunlight would not be able to penetrate the clouds of smoke and dust produced.

• Every day radiosonde balloons make over 1,000 measurements of temperature, winds, pressure, and humidity in the upper atmosphere.

• As a result of global warming, it has been estimated that world temperatures could rise by 4.5°C (40°F) by the year 2050. The resultant rise in sea level could be as much as 1.8 m (about 6 ft.) and is likely to cause flooding in low-lying areas. Places as diverse as Indonesia, the Netherlands, Bangladesh, and Central America risk flooding on a major scale, and often in areas of high population.

• The strongest wind ever measured on the surface of the Earth was 362 km/h (225 mph) on Mount Washington, New Hampshire, on April 12, 1934. Wind speeds can be even higher inside tornadoes.

• The Beaufort wind scale is used by seafarers and weather forecasters to indicate wind speed. It was devised in 1805 by the Irish hydrographer Francis Beaufort (1774–1857). It ranges from force 0 for speeds below 1.6 km/h (1 mph) to force 12 for speeds above 117 km/h (75 mph).

• Winds would be very different without the Coriolis Force, caused by the rotation of the planet. It is the same force that causes water to circle a drain in a consistent direction, depending on whether it is in the Northern or Southern Hemisphere.

• The prevailing westerly winds in latitudes 40°N and 40°S used to be called the "Roaring Forties" by sailors.

• It is possible to "seed" a cloud from aircraft with dry ice (carbon dioxide) pellets or salt particles in order to induce rainfall. It has also been found that seeding a cloud with silver iodide crystals can prevent heavy snow or a hailstorm.

• High-speed cameras show that a single flash of lightning may consist of as many as 42 "strokes" each lasting no more than 0.0002 seconds.

• In the United States, about 100 people are killed and many injured by lightning every year, more than by hurricanes or tornadoes. Roy Sullivan of Virginia was struck seven times between 1942 and 1977.

• The development of high-altitude airplanes has introduced artificial clouds known as contrails (condensation trails). These are formed from the condensed water vapor ejected as a part of the engines' exhaust gases.

Weather systems

The pattern of weather depends on the distribution of regions of warm air and cold air. This, in turn, depends on the distribution of low pressure areas and high pressure areas over the globe. When the Sun shines on the Earth's surface, the ground warms up. This warms the air above it. Warm air is lighter than cool air and so it rises, producing an area of low atmospheric pressure. In the cooler areas round about, the air is denser and at a higher pressure. It begins to move in toward the low-pressure area. This movement produces the winds. The global wind pattern can be pictured as a reflection of the areas of high pressure and low pressure across the world.

When air rises it cools. But cool air cannot hold as much moisture as warm air. In a cooling air mass, the moisture forms water droplets. These droplets in turn form clouds and, ultimately, fall to Earth as rain.

As a general rule, wherever there are rising air masses there is rain. The air may be made to rise by convection currents, by winds blowing up a mountain range, or by a cold air mass pushing its way below a warm air mass where the two meet along a "front." Such fronts are found between the cold air over the North Pole and the warmer air over the tropics. They account for the unstable weather patterns in temperate Europe and North America.

▼ A front is the boundary between two air masses. The boundary is not stationary, but moves across the landscape. If a warm air mass is replaced by a cold one, then it is a cold front, marked on a weather map by a line with teeth. A cold air mass is replaced by a warm one at a warm front, represented by a line with semicircles. At a cold front the cold air wedges its way

beneath the warm, forcing the warm air upward. Thunderclouds often form here as the warmer air is swept up quickly. At a warm front the warm air rises gently above the cold, producing a series of clouds at different heights. When one cold air mass catches up with another and the intervening warm air is lifted, the result is an occluded front.

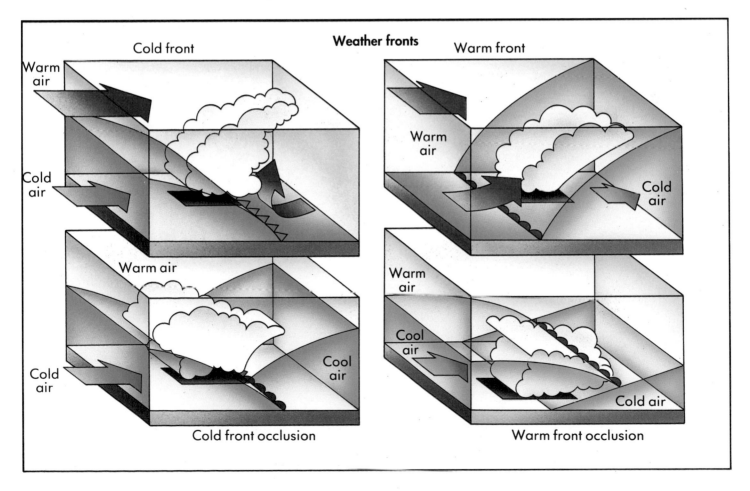

Wind

Winds are produced when cool air moves in to replace the warm air rising in a low-pressure area. The world distribution of high and low pressures determines the pattern of the usual, or prevailing, winds.

The Sun is always directly overhead somewhere between the tropics. As a result, the land areas along the Equator are among the hottest on Earth. The hot air rises there, resulting in an equatorial low-pressure belt. Air from the north and south sweeps in to equalize the pressure. This gives rise to the Northeast Trade Winds and the Southeast Trade Winds. (Winds are always named after the direction from which they blow.) The north-south movement of air is deflected toward the west as a result of the turning of the Earth.

The warm air that rises at the Equator spreads northward and southward at the top of the troposphere. There it cools before descending again in the regions of the Tropics of Cancer in the north and Capricorn in the south. Tropical high-pressure belts form there. The world's greatest deserts are found along these belts, because the descending air is dry. The air that descends may then return toward the Equator as the Trade Winds. Or it may spread toward the more temperate regions as the warm Southwesterlies in the Northern Hemisphere or the Northwesterlies in the Southern Hemisphere. These winds are usually referred to simply as the Westerlies.

Over the North and South Poles the cold temperatures give rise to high-pressure regions of cold air. Cold winds spread outward from these regions and meet the Westerlies along frontal systems in the temperate regions. Where they meet, the weather patterns are unstable. These major air movements produce the basic circulation pattern of the world's winds. It is disrupted and altered by the distribution of land and sea, and by the presence of mountain ranges. All of these factors decide the various climates of the world.

▼ Damage caused by a tornado in Pennsylvania in 1965. The tremendous power of the winds is spectacularly illustrated by the damage done by storms.

70

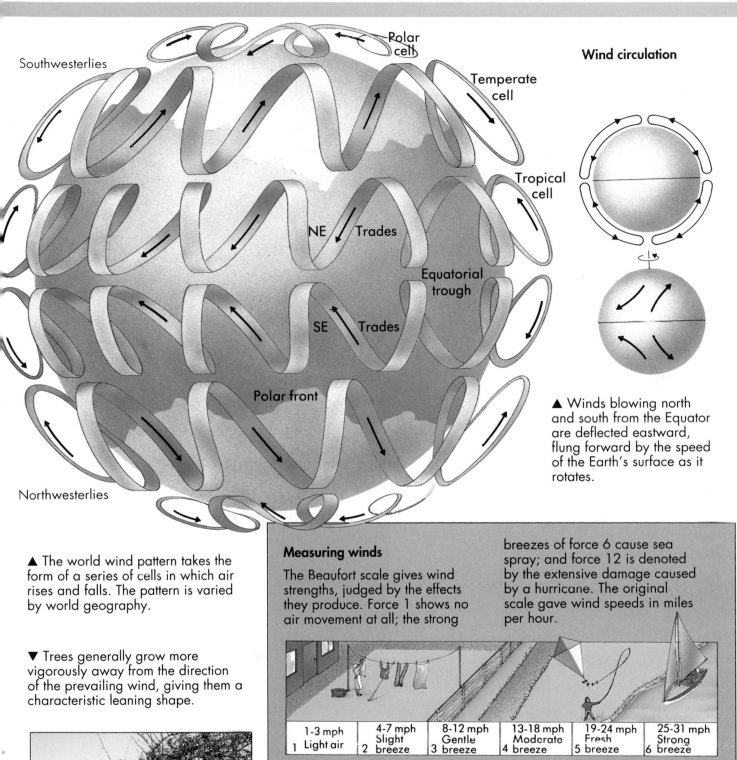

Southwesterlies

Polar cell

Temperate cell

Wind circulation

Tropical cell

NE Trades

Equatorial trough

SE Trades

Polar front

Northwesterlies

▲ Winds blowing north and south from the Equator are deflected eastward, flung forward by the speed of the Earth's surface as it rotates.

▲ The world wind pattern takes the form of a series of cells in which air rises and falls. The pattern is varied by world geography.

▼ Trees generally grow more vigorously away from the direction of the prevailing wind, giving them a characteristic leaning shape.

Measuring winds

The Beaufort scale gives wind strengths, judged by the effects they produce. Force 1 shows no air movement at all; the strong breezes of force 6 cause sea spray; and force 12 is denoted by the extensive damage caused by a hurricane. The original scale gave wind speeds in miles per hour.

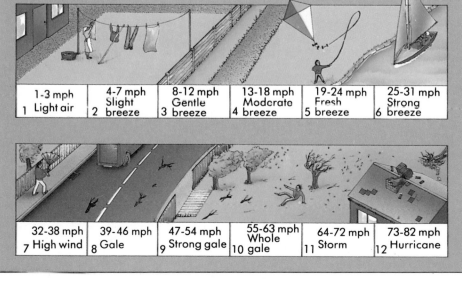

| 1 | 1-3 mph Light air | 2 | 4-7 mph Slight breeze | 3 | 8-12 mph Gentle breeze | 4 | 13-18 mph Moderate breeze | 5 | 19-24 mph Fresh breeze | 6 | 25-31 mph Strong breeze |

| 7 | 32-38 mph High wind | 8 | 39-46 mph Gale | 9 | 47-54 mph Strong gale | 10 | 55-63 mph Whole gale | 11 | 64-72 mph Storm | 12 | 73-82 mph Hurricane |

71

Measuring the weather

Weather forecasting has always been of great importance to people because of the weather's influence on shipping and navigation, farming, and almost every other aspect of human life. Before the development of scientific instruments, weather forecasting relied on observation. People tried to predict the weather by observing such things as wind direction, cloud types, sea color, and so on. Then in 1643 the Italian physicist Evangelista Torricelli invented the mercury barometer, which measures atmospheric pressure. From that time on, the study of weather became a very much more exact science, now called meteorology.

Modern meteorology contains a number of branches. "Dynamic meteorology" is the branch that deals with the movements of the atmosphere. It takes the sciences of hydrodynamics – the study of movements of liquids and gases – and thermodynamics – the study of the transfer of energy through a system – and applies them to the whole vast ocean of the air that surrounds us.

"Micrometeorology" is the branch of meteorology that deals with more localized effects, such as the development of land and sea breezes, and the formation of valley winds.

▼ Modern meteorologists use radar to help them track the movements of storms. Radar signals are reflected by raindrops and ice particles in clouds.

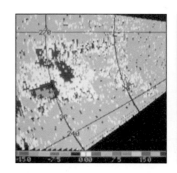
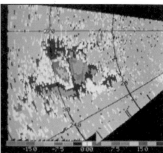

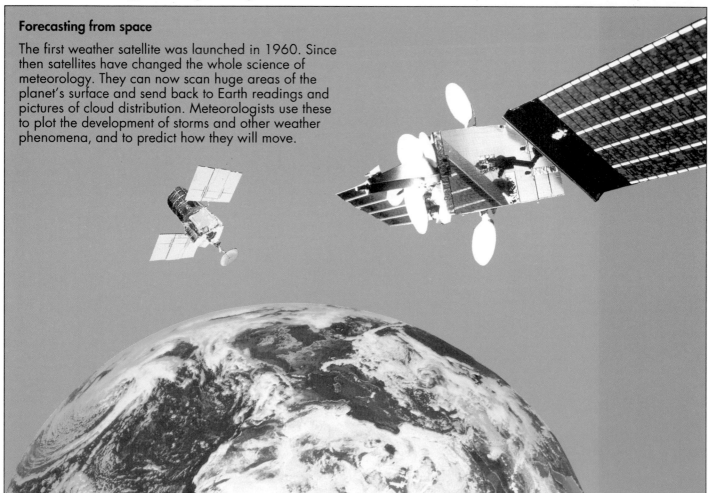

Forecasting from space

The first weather satellite was launched in 1960. Since then satellites have changed the whole science of meteorology. They can now scan huge areas of the planet's surface and send back to Earth readings and pictures of cloud distribution. Meteorologists use these to plot the development of storms and other weather phenomena, and to predict how they will move.

72

"Synoptic meteorology" is what we normally think of as weather forecasting. Readings of temperature, humidity, air pressure, cloud cover, wind strength and direction, and so on are taken at different places at the same time. This information is then plotted on a chart, or map, to give an overall view of the weather conditions at any particular time. Weather forecasters also use photographs of the clouds taken from orbiting satellites.

▼ A modern weather center. Information from a large number of weather stations is fed into computers, which produce synoptic charts detailing the weather conditions over a wide area. The charts are used to make weather forecasts.

Barometers

When the Torricelli barometer (below left) is up-ended in a reservoir of mercury, the level in the tube falls to approximately 760 mm (30 in.). With the aneroid barometer (below right), the needle moves round the scale according to the air pressure outside.

73

The changing climate

Over millions of years world climates change. The way they have changed can be seen by looking at rocks. In a particular area we might find beds of coal that were produced in a tropical swamp, covered by sandstones that formed in a desert. These may be covered, in turn, by mudstones deposited at the bottom of a shallow sea. Climate changes such as these take place over millions of years.

Extensive changes of climate can take place over shorter periods of time. The ice ages that began 2 million years ago – not a long time in geological terms – did not finish until 10,000 years ago. Throughout that time the world's climate varied widely. At times much of the Northern Hemisphere was choked with ice caps and glaciers. Then a few thousand years later the climates were warmer than they are now. A few thousand years later still the glaciers swept southward again.

Even in historical times there have been major changes in the climate. On the Tassili Plateau in the middle of the Sahara Desert, there are old rock paintings showing grassland animals. They must have been painted when the local climate was much moister than it is now. Trees still grow nearby. They have immensely long roots that extract water from

▲ The Mexican volcano El Chichón erupted in 1982, sending 16 million metric tons of dust into the atmosphere. The result was an enormous dust veil that absorbed some of the Sun's light, leading to a lowering of the Earth's surface temperatures.

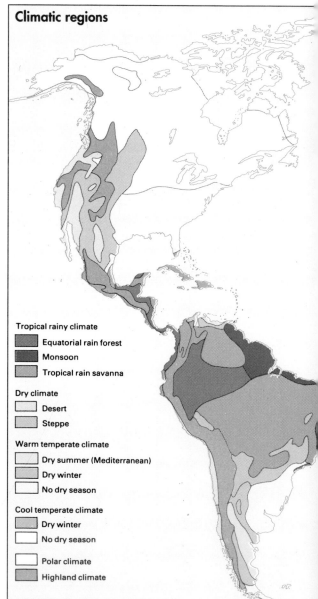

Climatic regions

Tropical rainy climate
- Equatorial rain forest
- Monsoon
- Tropical rain savanna

Dry climate
- Desert
- Steppe

Warm temperate climate
- Dry summer (Mediterranean)
- Dry winter
- No dry season

Cool temperate climate
- Dry winter
- No dry season

- Polar climate
- Highland climate

deep rocks. These trees could not have started growing unless there was water on the surface. During the thirteenth and fourteenth centuries, Europe suffered a "Little Ice Age." The climate was very much colder than it is now. In London, for example, winter fairs were often held on the frozen River Thames. This would be impossible nowadays because it does not get cold enough.

The changes in climate through geological time, as revealed by the different rocks, can be explained by the drifting of the continents from one climatic region to another. More recent changes are due to shorter-term events. Volcanic eruptions can throw up dust and gases such as sulfur dioxide high into the atmosphere. There they can block out sunlight and lower the temperatures on the Earth's surface. A noticeable cooling in the 1960s coincided with increasing volcanic activity across the globe.

Another influence may be a fluctuation in the energy output of the Sun itself. Old astronomical records show that the Sun does, indeed, change in size and energy output from time to time. These changes alter the climate.

▼ The Earth can be divided into a number of distinct climatic zones or biomes. Changing conditions may mean that maps like this will be inaccurate within a few decades.

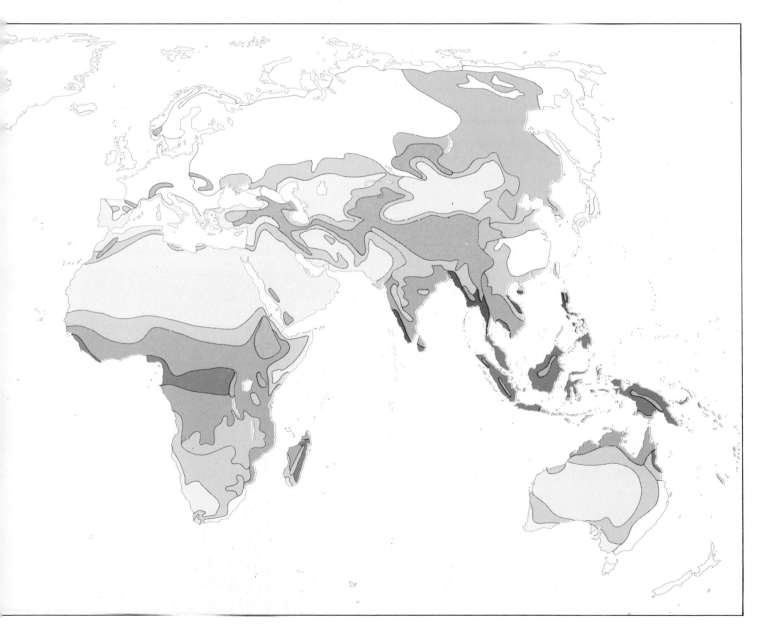

Climates of the world

The Sun's heating effect is stronger at the Equator than at the poles. This produces different surface temperatures over the face of the Earth. It also produces areas of high and low atmospheric pressure, and these generate the pattern of prevailing winds. The distribution of land and sea, the patchwork of the continents, and the sweep of the mountain ranges modify the pattern produced. The result is a huge range of climates and environmental conditions in different parts of the world.

SPOT FACTS

- The coldest temperature recorded for North America was -65°C (-85°F) at Snag in the Yukon.

- The coldest place on Earth is the Pole of Inaccessibility in Antarctica, with an average annual temperature of -58°C (-72°F).

- The hottest place in the United States is Death Valley, California, with temperatures of up to 57°C (135°F).

- The world's most disastrous flood was in 1887 when the Hwang-ho (Yellow River) in China burst its banks, killing 900,000 people.

- The rain forest has abundant rainfall all year, totaling 200–400 cm (80–160 in.) annually.

- There are more species of plants and animals in tropical rain forests than all the rest of the world's ecosystems put together.

- In deserts, daytime temperatures often reach 55°C (131°F). At night the desert floor radiates heat back into the atmosphere, and the temperature drops to near freezing.

- There is always snow on Mount Kilimanjaro, even though it is on the Equator.

- Grassland soils are fairly fertile. Due to low rainfall, nutrients are not lost and due to the absence of trees, the grasses are able to build up a rich topsoil. In northern temperate regions, grasslands are used for grain crops such as corn (maize) and wheat. The native grazing animals have been ousted by domestic cattle and sheep, and the large predators have been destroyed.

- Mounds of sand built by the wind may reach heights of more than 200 m (more than 650 ft.) in the Sahara, Arabian, and Iranian Deserts.

- A growing sand dune may travel as much as 30 m (100 ft.) in one year.

- Groundwater can be tapped from aquifers underlying deserts, but the water is fossil water, which, once used, is not replaced. Artificial oases have been made in the Sahara by drilling wells more than 1,000 m (more than 3,280 ft.) deep.

- The Sahara, the largest desert in the world, is about 1,610 km (about 1,000 mi.) wide and about 5,150 km (about 3,200 mi.) long from east to west. The total area is more than 9,065,000 sq km (more than 3,500,000 sq. mi.).

- Several mountain peaks in the Sahara Desert are more than 3,414 m (more than 11,200 ft.) high. A number of the central Saharan peaks are snowcapped for part of the year.

- The Arctic has more than 400 species of flowering plants.

- Areas in the center of large continents have more extreme climates because of their distance from the sea. Milwaukee, Wisconsin, is on approximately the same latitude as Rome in Italy. However, Milwaukee frequently records winter temperatures of below -17.7°C (0°F) because of its continental climate.

- The greatest range of temperature recorded for one day was at Browning, Montana, in January 1916. The temperature fell from 7°C (44°F) to -49°C (-56°F). At the other end of the scale, the temperature on one day in the Sahara Desert has ranged from 52°C (126°F) to -2°C (28°F).

Rain forest climate

The Sun is almost overhead at the Equator. Its rays come vertically downward, cutting straight through the atmosphere and concentrating their energy on small areas of ground. The air becomes very hot and it rises, producing a belt of low pressure along the Equator. This draws in the Trade Winds from the northeast and southeast. The winds usually travel over oceans and so their air is very moist. When the air reaches the low-pressure areas, it begins to rise. Clouds form and drop their water on the land beneath as daily torrential downpours.

This vigorous circulation of water produces vast networks of streams flowing into the greatest rivers of the world. The Amazon, the Zaire, and the Mekong all flow close to the Equator. The humid lowland plains they flow through have a hothouse atmosphere. Plants grow in profusion and produce the tropical rain forest. The conditions are so good for plant life that many thousands of different species can exist in a few square miles. They produce trees up to 70 m (230 ft.) tall, all growing past each other to reach the sunlight.

Smaller plants called epiphytes, or air plants, grow on the branches, and others in the form of creepers climb up the trunks to reach the light. The intertwined and entangled mass of branches, leaves, and creepers forms a green canopy over the whole forest. The tallest trees, the emergents, reach through the canopy into the air above. The forest floor is dark and hot, and few plants grow except where a fallen tree leaves space for the Sun to shine in. Along the river banks the crowns of branches and leaves come right down to ground level.

The vast range of plant types supports a variety of animals as well. Most of these stay among the sunlit branches, although some live in the darkness below.

◀ The gibbon of southeastern Asia is the most agile of tree-dwelling animals and it eats fruits. The wide variety of plants in the tropical forest has led to the evolution of a range of animals that feed on them.

▼ The several species of sloths in South America are among the slowest moving of the tree-dwelling mammals. They hang upside down and live on a diet of leaves.

Grasslands

Grasslands develop naturally in regions where the climate is generally dry but has distinct moist seasons. Grasses are able to weather long periods of drought because of their underground stems. The leaves and heads may die off in dry weather, but they can grow again from the underground part of the plant. Trees generally do not thrive in such conditions, and the typical landscape is one of wide open plains with very few trees.

Tropical grasslands, or savannas, occur in two bands north and south of the Equator. They lie between the central belt of equatorial rain forest and the two belts of desert along the tropics. As the Earth moves round in its orbit and the Sun appears to move north and south in the sky between summer and winter, the rainy conditions that produce the tropical forest move north and south too. The regions that lie between the forest and desert belts get both types of climate. They have forest weather at one part of the year, and desert weather at another. The resulting vegetation, the tropical grassland of savannas, can be traced from one continent to another, but savannas are most prominent in Africa.

▼ The grasslands that we see in moist, temperate regions tend not to be natural. Throughout civilization communities have chopped down trees and cut clearings for fields. These take on the appearance of grasslands because the important cereal crops are themselves grasses.

▶ Grassland animals, such as these pronghorn antelopes, have very strong teeth for chewing grass, and complex digestive systems for digesting it. They also have long legs and are built for speed. The best method of defense against predators is to run away.

Temperate grasslands are found deep within continents, usually bordering on cool desert areas. The prairies of North America and the steppes of Asia are northern examples. In the south, the pampas of South America are partly temperate and partly tropical.

Grassland animals are highly specialized, because grass is a difficult food to digest. The development of grasslands 50 million years ago allowed grazing animals such as antelopes, horses, and cattle to evolve. In turn, grazing stimulates fresh growth and the animal dung keeps the ground fertilized.

◀ Elephants are not typical of the grassland animals. They are slow moving, and rather than feeding on grass, they tend to eat the shoots and leaves of the few bushes and trees that grow there.

▲ The South American rheas, the African ostrich, and the Australian emu share the same grassland lifestyle. Flightless birds are typical of grasslands. Their long legs enable them to run quickly over the plains.

◀ The long face of the zebra is typical of a grassland animal. Its mouth can be munching the grass while its eyes are still quite high and looking about. Zebras graze in herds, with one or two animals always looking out for danger.

79

Deciduous and boreal forest

Deciduous trees lose their leaves in winter and grow new ones each spring. The deciduous woodlands are found in the temperate regions of the Earth, mostly in the Northern Hemisphere. The temperate zone is a broad band which, at times, is subjected to the cold wind coming from the direction of the poles. At other times, it is exposed to the warm Westerlies. As a result, the climate is a mild one compared to other climates of the world. It is generally moist and does not have extremes of heat or cold.

This zone has particularly favorable places for people to live and grow crops. Over the centuries, much of the original deciduous forest has been cleared away for cities and farms. The common large trees of these woodlands are broad-leaved types, such as oaks, ashes, beeches, and willows. Smaller trees growing beneath them include maples and birches. At a lower level still grow the bushes of dogwoods, hollies, and hawthorns, and there is usually a thick undergrowth of flowering plants.

To the north of the belt of deciduous woodland lies the largest stretch of uninterrupted forest in the world. The great coniferous boreal forest stretches from Scandinavia eastward across northern Europe and Asia, then across Alaska and Canada. There the prevailing weather is brought by the cold air masses that blow in from the far north. The growing season is only three or four months long, and during the lengthy winter all the moisture is locked up in ice and snow. Coniferous trees are able to withstand these conditions.

▶ Deer are typical animals of deciduous woodland. They eat a number of different foods, including leaves and young shoots from the trees and the undergrowth.

▼ A deciduous woodland has several different kinds of trees, along with bushes and an undergrowth of many small flowering plants.

The needle shape of coniferous tree leaves reduces the rate at which water is given off through them. The leaves stay on the tree all year round, so that they are ready as soon as the growing season begins. The trees' conical shape allows snow to slide off easily.

Although coniferous trees are particularly well adapted for the cold conditions, they are found in more temperate regions as well. The deciduous woodlands rarely consist of only deciduous trees but usually have conifers among them – giving a "mixed woodland."

▲ There are fewer animals and birds in coniferous forests than in deciduous woodlands, and they tend to be quite specialized feeders. Crossbills (1), for example, eat only seed cones. The coniferous forests of Canada (2) are home to the beaver – a creature that can alter its habitat. Beaver colonies can fell trees and dam rivers, holding back lakes in which they build their lodges. The blunt face of the beaver (3) hides a massive set of chisel-like incisor teeth, which it uses to fell the trees. Groups of about 12 animals cooperate in building the dams and lodges.

Mountain climate

Conditions change as we climb a high mountain. The higher we go, the thinner the air becomes. The sky also gets bluer and the Sun's rays become hotter. There is a change in climate between the mountain's base and its summit. The difference can be similar to the difference between the climate at the Equator and at the North Pole or South Pole.

At the base of an equatorial mountain, such as Mount Kenya or Mount Kilimanjaro in eastern Africa, there may be tropical forest or tropical grassland. Higher up the temperature falls, and the climate becomes moister because of rain falling from rising clouds. At about 1,500 m (5,000 ft.) the original tropical conditions give way to those of moist temperate forests.

Above about 2,400 m (8,000 ft.) there is less rainfall and the forest gives way to scrub. Grasses become the main plants. In the African and Asian mountains, bamboo is common.

The bamboo and scrub give way at a height of about 3,000 m (10,000 ft.) to Alpine meadows, with coarse grasses and heathers. Up there the conditions are too harsh for trees to grow, and the landscape is similar to that of the tundra regions of the far north.

Above the tree line
Just as the tundra in the north gives way to the ice caps of the North Pole, so the Alpine meadows of the mountain give way to the glaciers and snow-covered crags of the mountain summit. On top of the African mountains there are permanent snowfields and glaciers, even though they lie near the Equator.

Most mountain slopes are gentle, especially among the foothills. Changes in climate usually take place over quite large distances. But in the Himalayas, the gorge of the Brahmaputra River is so narrow and steep that conditions change from tropical forest to snow and ice within a mile or two.

▶ Bighorn sheep in Alaska. The change in climate between the base of a mountain and its summit is accompanied by a change in plant and animal life. Typical forest animals such as hogs and deer give way to more specialized creatures on the higher slopes. In Asia, giant pandas live at this level, and in Africa there are mountain gorillas. Surefooted mountain goats and ibexes live on the sparse grazing of the higher reaches.

82

Ice

Tundra

Scrub

Cloud forest

Rainforest

Arid ground

Deserts and polar climates

Plants need a certain minimum amount of water to survive. Very few plants can grow in particularly dry regions, where there is very little rainfall or all the moisture is locked up as ice. The barren landscape that results is known as a desert.

Hot deserts are the kind that usually come to mind when we think of a desert. There are several types of hot deserts. Tropical deserts lie in two belts along the Tropics of Cancer and Capricorn. Hot air that has risen and released its moisture as rain over the low-pressure equatorial belt now descends over the tropics. This dry air forms high-pressure belts and no moist winds blow over these areas. The great deserts of the Sahara and Arabia lie in this zone in the north. In the south there are the Kalahari in Africa and the Gibson in Australia.

Continental deserts exist in places that are so far from the sea that the moist winds just cannot reach them. The Gobi Desert in central Asia is a typical example.

Finally there is the rain shadow desert, which is found on the lee side of mountain ranges (the side away from the wind). Winds from the ocean lose all their moisture in rain as they rise up the seaward mountain slopes, leaving only dry air to pass over to the other side. California's Death Valley is the most famous rain shadow desert.

Hot deserts are usually surrounded by zones of semidesert, in which only specialized types of plants can grow. This is also true of the cold deserts. Beyond the northern reaches of the great coniferous forests there is a region in which the climate is too harsh for trees. It is an area of permanently frozen subsoil with a vegetation of coarse grasses, heathers and

▲ Cactus plants have a thick fleshy stem that holds water, a leathery waterproof skin to keep the moisture in, and spines to protect them from damage by animals. All these features enable them to survive in harsh desert conditions.

◀ Death Valley in California is the hottest place on Earth. For most of the year it is totally dry. But when the rains do come they occur as torrential downpours with several inches falling in a day. Immediately afterward the desert blooms with all kinds of plants before it returns to barrenness for the remainder of the year.

84

other small herbaceous plants (plants lacking a woody stem). This is what is called the tundra; the word "muskeg" is used in Canada.

Islands close to the North Pole and the continent of Antarctica at the South Pole can be regarded as true deserts. As a result of the Earth's tilt, the Sun shines there for only part of the year, and when it does its rays slant through thick layers of atmosphere. Because of the permanently low temperatures water is always frozen and useless to living things.

▼ The icy lands of the far north are barren enough to be considered deserts. However, the sea supports a great deal of life. Algae grow in the waters, and are eaten by fish. The fish are food for seals that are, in turn, hunted by polar bears. Birds, such as gulls, act as scavengers.

▲ Small pillow-like plants grow all over the tundra. Their compact shape gives them protection against the severe weather.

▶ Reindeer graze on the tundra vegetation in summer, but migrate southward to the forest in winter.

85

Units of measurement

Units of measurement

This encyclopedia gives measurements in metric units, which are commonly used in science. Approximate equivalents in traditional American units, sometimes called U.S. customary units, are also given in the text, in parentheses.

Some common metric and U.S. units

Here are some equivalents, accurate to parts per million. For many practical purposes rougher equivalents may be adequate, especially when the quantity being converted from one system to the other is known with an accuracy of just one or two digits. Equivalents marked with an asterisk (*) are exact.

Volume
1 cubic centimeter = 0.0610237 cubic inch
1 cubic meter = 35.3147 cubic feet
1 cubic meter = 1.30795 cubic yards
1 cubic kilometer = 0.239913 cubic mile

1 cubic inch = 16.3871 cubic centimeters
1 cubic foot = 0.0283168 cubic meter
1 cubic yard = 0.764555 cubic meter

Liquid measure
1 milliliter = 0.0338140 fluidounce
1 liter = 1.05669 quarts

1 fluidounce = 29.5735 milliliters
1 quart = 0.946353 liter

Mass and weight
1 gram = 0.0352740 ounce
1 kilogram = 2.20462 pounds
1 metric ton (tonne) = 1.10231 tons

1 ounce = 28.3495 grams
1 pound = 0.453592 kilogram
1 short ton = 0.907185 metric ton (tonne)

Length
1 millimeter = 0.0393701 inch
1 centimeter = 0.393701 inch
1 meter = 3.28084 feet
1 meter = 1.09361 yards
1 kilometer = 0.621371 mile

1 inch = 2.54* centimeters
1 foot = 0.3048* meter
1 yard = 0.9144* meter
1 mile = 1.60934 kilometers

Area
1 square centimeter = 0.155000 square inch
1 square meter = 10.7639 square feet
1 square meter = 1.19599 square yards
1 square kilometer = 0.386102 square mile

1 square inch = 6.4516* square centimeters
1 square foot = 0.0929030 square meter
1 square yard = 0.836127 square meter
1 square mile = 2.58999 square kilometers

1 hectare = 2.47105 acres
1 acre = 0.404686 hectare

Temperature conversions

To convert temperatures in degrees Celsius to temperatures in degrees Fahrenheit, or vice versa, use these formulas:

Celsius Temperature = (Fahrenheit Temperature − 32) × 5/9
Fahrenheit Temperature = (Celsius Temperature × 9/5) + 32

Numbers and abbreviations

Numbers

Scientific measurements sometimes involve extremely large numbers. Scientists often express large numbers in a concise "exponential" form using powers of 10. The number one billion, or 1,000,000,000, if written in this form, would be 10^9; three billion, or 3,000,000,000, would be 3×10^9. The "exponent" 9 tells you that there are nine zeros following the 3. More complicated numbers can be written in this way by using decimals; for example, 3.756×10^9 is the same as 3,756,000,000.

Very small numbers – numbers close to zero – can be written in exponential form with a minus sign on the exponent. For example, one-billionth, which is 1/1,000,000,000 or 0.000000001, would be 10^{-9}. Here, the 9 in the exponent -9 tells you that, in the decimal form of the number, the 1 is in the ninth place to the right of the decimal point. Three-billionths, or 3/1,000,000,000, would be 3×10^{-9}; accordingly, 3.756×10^{-9} would mean 0.000000003756 (or 3.756/1,000,000,000).

Here are the American names of some powers of ten, and how they are written in numerals:

1 million (10^6)	1,000,000
1 billion (10^9)	1,000,000,000
1 trillion (10^{12})	1,000,000,000,000
1 quadrillion (10^{15})	1,000,000,000,000,000
1 quintillion (10^{18})	1,000,000,000,000,000,000
1 sextillion (10^{21})	1,000,000,000,000,000,000,000
1 septillion (10^{24})	1,000,000,000,000,000,000,000,000

Principal abbreviations used in the encyclopedia

°C	degrees Celsius
cc	cubic centimeter
cm	centimeter
cu.	cubic
d	days
°F	degrees Fahrenheit
fl. oz.	fluidounce
fps	feet per second
ft.	foot
g	gram
h	hour
Hz	hertz
in.	inch
K	kelvin (degree temperature)
kg	kilogram
l	liter
lb.	pound
m	meter
mi.	mile
ml	milliliter
mm	millimeter
mph	miles per hour
mps	miles per second
mya	millions of years ago
N	north
oz.	ounce
qt.	quart
s	second
S	south
sq.	square
V	volt
y	year
yd.	yard

Glossary

acid rain Rain that contains so much dissolved impurity that it gives an acid reaction. The impurities are usually oxides of sulfur and nitrogen from the burning of fossil fuels. The acidity can cause serious damage to trees.

acidic In geological terms, an igneous rock that contains a high proportion of silica. It has nothing to do with acidity in a chemical sense.

ammonite An extinct animal that resembled an octopus in a coiled shell. Fossils of ammonites are common in marine sedimentary rocks of the Mesozoic era.

anaerobic Referring to a reaction that takes place without oxygen.

aphelion The point in a planet's orbit at which it is closest to the Sun.

aquifer A water-bearing bed of rock.

artesian A well or a spring from which water gushes up under its own pressure.

asthenosphere A soft layer within the Earth's mantle on which the solid lithosphere moves.

atmosphere The envelope of gas that surrounds a planet. The Earth's atmosphere consists of nitrogen, oxygen, and small proportions of other gases.

aurora A light display in the sky caused by particles from the Sun's radiation reacting with the Earth's magnetic field. Usually called the Aurora Borealis, or Northern Lights, in the Northern Hemisphere, and the Aurora Australis, or Southern Lights, in the Southern Hemisphere.

axis The imaginary line through the Earth, or any other body, around which it spins.

barometer An instrument for measuring atmospheric pressure.

basic In geological terms, a rock that contains a low proportion of silica.

biochemistry The chemical reactions that take place inside a living body.

biogenic A term that describes a form of sedimentary rock. Biogenic sedimentary rock is rock formed from fragments of once-living matter, such as shells or wood.

block mountain An upland bounded by faults along which it was raised, or along which the surrounding landscape has subsided.

Cenozoic The era of geological time spanning from 65 million years ago up until the present day.

chemical A term that describes a form of sedimentary rock. Chemical sedimentary rock is rock formed from a solution, such as salt.

cirque A bowl-shaped hollow from which a glacier flows. It is widened and deepened by the weight of the glacier itself. When the glacier melts, it often leaves a cirque lake.

clastic A term that describes a form of sedimentary rock. Clastic sedimentary rock is formed from fragments worn off other rocks.

climate The sum of the atmospheric conditions at a particular place averaged out over a long period of time.

continent A large continuous landmass that rises above the ocean floor. Most geologists accept that there are seven continents. Asia, Europe, Africa, South America, North America, Australia, and Antarctica.

continental drift The phenomenon in which the continents appear to have moved throughout geological time. Such movements are now accounted for by plate tectonics.

convection A current that is formed as warm fluid, being lighter than its surroundings, rises. At the same time cooler fluid descends.

core The innermost part of the Earth's structure.

Coriolis effect An effect caused by the Earth's rotation, whereby anything moving toward the Equator is deflected to the west, and anything moving away from the Equator is deflected to the east.

crust The outermost part of the Earth's structure.

crystal A regular, naturally formed shape, usually with flat faces and sharp edges. Most substances will form crystals, the shape of which reflects their chemical structure.

cumulus A cloud with a heaped and hummocky-shaped appearance.

deciduous A term that describes a type of tree. A deciduous tree is one that loses its leaves in winter.

delta A structure of sandbanks, islands, and channels formed at the mouth of a river, where tides and currents cannot carry away the river's sediment.

dike An igneous intrusion consisting of rock cutting across the bedding of the surrounding rock. A dike is formed as a crack fills up with magma.

ecosystem A community of organisms all interacting with one another.

environment The sum total of conditions under which an organism lives. These conditions include such things as the climate, the soil, the water, and the other organisms present.

epicenter The point on the Earth's surface directly above the focus of an earthquake.

erosion The process whereby the rocks and soils of the surface of the Earth are broken down by the action

88

of the weather, by streams, by glaciers, or by human interference.

evolve Of a living organism, to change over generations in response to changing conditions, producing new species.

extrusive Term that describes a form of igneous rock. Extrusive igneous rock breaks through the Earth's surface before it solidifies.

fault A crack in the Earth's crust along which displacement has occurred, caused by tectonic movements.

focus The point at which most movement occurs in an earthquake.

fold A bend in rock caused by tectonic movements.

fold mountain A mountain formed by folding, usually as tectonic plates move against one another.

fossil The remains or trace of a once-living organism preserved in rock.

front In meteorology, the boundary between two air masses of different temperature. Fronts are usually associated with unstable weather.

geology The study of the Earth, its rocks and minerals, its fossils, and its changing conditions throughout time.

glacier A mass of ice, formed from compacted snow, that moves by its own weight down a valley.

gravity A force of attraction that exists between objects because of their mass. The most obvious result of this is that objects fall to Earth.

greenhouse effect A condition of the atmosphere that arises because certain gases that are present will allow radiation from the Sun to reach the Earth's surface, but will not allow the reradiated heat from the Earth's surface to escape,

thereby increasing the temperature. The artificial increase of such gases as carbon dioxide and water vapor from industry is a current cause.

groundwater Water that is held in the rocks and soil of the Earth.

gyre A large circular motion of seawater caused by ocean currents and the Coriolis effect. A gyre typically occupies half an ocean.

hard water Water that contains a high concentration of dissolved calcium carbonate. Such water produces mineral deposits in kettles and pipes.

historical geology That aspect of geology that describes the conditions in past times from analysis of the rocks formed at that time.

ice age A period of time in which the climate is colder than normal and glaciers are particularly widespread. The most famous is that which took place in the last 2 million years, but there have been several throughout geological time.

iceberg A mass of ice floating in the sea, broken off from a glacier or an ice floe.

igneous rock One that has formed as molten material, and has cooled and solidified, either on the Earth's surface (a volcanic or extrusive rock) or underground (an intrusive rock).

inselberg A rounded mass of rock protruding above an arid plain. It is a result of onion-skin weathering.

intrusion A body of igneous rock that has formed underground.

invertebrate An animal with no backbone. Effectively this is almost any animal that is not a fish, amphibian, reptile, bird, or mammal.

irrigation Bringing water to a dry landscape to enable plants to grow.

laccolith An intrusion that forces the sedimentary rock above

it to form a dome shape.

lava Molten rocky material that erupts from a volcano and solidifies on the Earth's surface.

lithosphere The solid portion of the Earth's surface. It consists of the crust and the topmost part of the mantle. The tectonic plates of the Earth's surface are sections of the lithosphere.

magma Molten rocky material beneath the Earth's surface.

magnetism A property, showing up most strongly in certain metals, that causes attraction and repulsion. The field (the space in which the effect is felt) has two ends, called the north pole and the south pole. When two magnetic objects are brought together, the unlike poles attract one another and the like poles repel. An electric current can produce a magnetic effect.

mantle The stony section of the Earth between the crust and the core.

meander A loop in a river's course, typical of a river's old age.

Mesozoic The era in geological time that lasted between 248 and 65 million years ago.

metamorphic rock Rock that has been heated and/or crushed so much by the movements of the Earth that the components have recrystallized into new minerals.

meteorite A body of rock or particle of dust from space that reaches the ground.

meteorology The study of the movement of the atmosphere and the weather.

mid-ocean ridge An elevated part of the ocean bed that winds throughout the oceans of the world. It represents a tectonic plate boundary, where new material is welling up from the Earth's interior.

mineral A naturally formed inorganic substance with a consistent chemical composition. All rocks are made up of an aggregation of minerals.

monsoon A seasonal effect in the countries bordering the north Indian Ocean, in which cold winds blow outward from the Asian continent in the cooler times of the year, and warm, wet winds blow inward in the warmer times.

moraine An accumulation of rocky debris that has been scraped up and carried along by a glacier.

nebula A cloud of gas and dust in space.

oasis A moist, fertile spot in a desert area.

observatory An institution or building where natural phenomena, such as the stars, earthquakes, or the weather, are studied.

ocean trench An elongated deep section of the ocean floor, usually found close to mountainous continents or island arcs. It represents a tectonic plate boundary, where one old plate is being dragged down beneath another.

oceanic zone The area of open sea lying beyond the edge of the continental shelf and where the depth is greater than 200 meters (650 ft.).

oceanography The study of the oceans.

ooze Deep-ocean sediment formed largely from the remains of living creatures.

orbit The path in space of one body around another, as the Moon around the Earth. The forward momentum of the first body just balances the gravitational effect of the second, so that the one falls around the other without touching it.

ore mineral A mineral that contains a valuable metal which can be extracted economically.

organic Referring to the chemistry of living things.

oxbow A curved lake left behind as a meander is cut off when a river changes its course.

ozone A type of oxygen that contains three atoms of oxygen in its molecules instead of the usual two.

ozone layer A layer in the atmosphere where ozone is being generated. This acts as a shield against ultraviolet solar radiation. Deterioration of the ozone layer, through pollution of the atmosphere, will allow more harmful radiation to reach the Earth's surface.

Paleozoic The era in geological time between 590 and 248 million years ago.

perihelion The point in a planet's orbit at which it is farthest from the Sun.

planet A large body in orbit around the Sun or a similar body in orbit around another star.

plate tectonics The movement of the Earth's surface, caused by the generation of new plate material at the mid-ocean ridges, its spreading, and its destruction in ocean trenches. This worldwide movement takes in the old concept of continental drift and the newer concept of seafloor spreading.

polarized light Light in which the waves vibrate in one plane only. This can be the result of reflection of the light from a flat surface, or of the light's passage through a polarizing filter. Polarized light can be blocked by filters that are polarized at right angles to the beam. Certain minerals distort polarized light to produce colors that can be used to identify them.

pollution The poisoning of an

environment by unwanted substances.

prevailing wind A wind that blows from a particular direction for most of the time.

reef A rocky outcrop that produces a shallow area out at sea. The term is used especially for the structures built by the action of corals.

regional A term that describes a form of metamorphic rock. Regional metamorphic rock is rock formed by tectonic pressure.

rift valley A valley produced as an area of rock subsides between two faults, or folds downward against a single fault.

rock In geological terms a rock is any substance that makes up the surface of the Earth, and includes consolidated sediments. More commonly it refers to solid matter formed of minerals that have grown in a mass or have become cemented together.

rock cycle The process by which a rock forms. Rock may be lifted up by tectonic forces, then broken down by erosion, with the resulting fragments being consolidated into a new rock. Or alternatively the original rock melts and solidifies into a new rock.

salinity A measure of the amount of solid substances dissolved in seawater, that is, its saltiness.

seafloor spreading The phenomenon whereby the ocean floor is observed to become older farther away from the mid-ocean ridges. The discovery was made in the 1960s and is now covered by the all-embracing concept of plate tectonics.

sediment Any loose material deposited in layers by water, ice, or wind.

sedimentary rock One that has formed as loose sediments have

90

become compacted and cemented together.

seismograph A device for detecting and measuring earthquakes.

serac A pinnacle of ice formed on a glacier where two sets of crevasses intersect.

sial Scientific shorthand for the substance of the continental crust. It is an abbreviation of the major constituents, silicon and aluminum.

silicate A mineral that contains silicon, oxygen, and usually some other element. Silicates are the most common rock-forming minerals.

sill An igneous intrusion, consisting of a sheet of rock lying parallel to the bedding of the other rocks of the area.

sima Scientific shorthand for the substance of the oceanic crust. It is an abbreviation of the major constituents, silicon and magnesium.

soil A loose covering of natural material, consisting of broken-down rocky material mixed with decomposing plant matter.

Solar System The group of bodies in space comprising the Sun, along with the planets, all their moons, and an unknown number of rocky bodies and comets, all under the gravitational influence of the Sun.

stalactite A hanging structure in a cave formed by the buildup of calcite from drops of water on the cave roof.

stalagmite A buildup of calcite on a cave floor, formed by the calcite coming out of solution from drops of water hitting the floor.

stratus A cloud with a layered appearance.

subduction zone A region where the edge of one tectonic plate is being destroyed beneath that of another. Subduction zones are

usually marked by the presence of ocean trenches.

subsoil The layer of soil between the topsoil and the bedrock, containing little organic matter.

survey To study a tract of land by making measurements and investigating its features.

technology The application of science to industry, research, commerce, and other fields of human endeavor.

tectonics The study of the forces that move rocks, the movements themselves, and the structures, such as faults and folds, that are so formed.

thermal A term used to describe a form of metamorphic rock. Thermal metamorphic rock is formed by very high temperatures.

tornado A rapidly spinning column of air, caused by an intense convection current usually along a front. Sometimes called a whirlwind.

Trade Winds Prevailing winds that blow toward the Equator from the northeast and the southeast. So called because they were vital to the trade routes before the advent of powered ships.

tropic One of the two lines of latitude that mark the boundaries of the zone in which the Sun is overhead sometime during the year. The Tropic of Cancer is $23\frac{1}{2}°$ N. The Tropic of Capricorn is $23\frac{1}{2}°$ S.

troposphere The lowest layer of the Earth's atmosphere, up to a height of about 11 km (7 mi.). All weather phenomena occur and all life exists within this layer. The upper boundary is irregular, being higher in the tropics than at the poles.

tundra The landscape of the far north, characterized by a lack of trees and a very wet appearance in the summer. The significant feature is the permafrost – a layer of

permanently frozen subsoil that prevents the summer meltwater from draining away.

typhoon The term used in the Far East for a hurricane.

ultraviolet Any wavelength of light that is too short to be seen by the naked eye.

vent A fissure or a tunnel in a volcano, through which lava or gas emerges.

volcano A vent in the Earth's surface, from which lava and gas are expelled from the interior. The solidified lava around the vent usually builds up into a mountain.

water table The upper boundary of the underground zone in which the rocks are saturated with water.

weather The daily changing climatic conditions at any place on the Earth's surface.

weathering Erosion caused by the influence of weather conditions. Such conditions include rain, freezing water, sandblasting by wind, heating by the Sun, or chemical decomposition by acids dissolved in rainwater.

91

Index

Page numbers in *italics* refer to pictures. Users of this Index should note that explanations of many scientific terms can be found in the Glossary.

A

aa *32*
ablation moraine *42*
abyssal plain 53, *53*
accretion theory *10*
acid rain 62, 67
aftershocks 26
air 63
    composition of *63*
Alaska 80
algae *66*
altimeter 62
Amazon River 38, 49, 77
ammonites *29*
anaerobic *66*
Andes 19, 23
andesite 33
Antarctic Ocean 50, *50*
Antarctica 76, 85
antelope 79
aphelion 12
aquifer *59*
Arabian Desert 84
Arctic Ocean 50, *50*
arête *42*
argon 62, 63, 66
artesian *59*
ash (tree) 80
Asia 80
asthenosphere 14, *15*, 16, *17*
atmosphere 60, 62-67
    composition of *64, 65, 67*
    saturated 60
    structure of 64
atmospheric pressure 64, 69, 70
atoll 52
aurora *65*
Australian Current 50
avalanche *42*

B

bacteria *66*
bamboo 82
barometers *73*
    see Toricelli
basalt *see rock*
Beaufort scale 71, *71*
beaver *81*

beech 80
bergschrund *42*
Bering Strait 50
Bighorn sheep *82*
Bingham Canyon Copper Mine *44*
biomes *74-75*
birch 80
block mountains 23
borehole 57, *59*
Brahmaputra River 82
broad-leaved tree 80

C

cactus *84*
calcite 39, 66
calcium carbonate 54
California Current 50
Canada 80
carbon dioxide 9, 62, 63, 66, 67
carbon monoxide 66
chalk *35*
Cheddar Gorge *34*
chlorine 49
cirque *42*
climate 68-70, 74-77, *74-75*, 80
cloud 54, 60, *60,* 69
    cirrus *60*
    cumulonimbus *60*
    cumulus 60
    heaped 60
    stratus 60
    stratocumulus *60*
coal *34-35*
cold desert 84
conglomerate *see rock*
coniferous trees 80-81
constructive plate margins *17*, 23, 24
continental crust 52, *53*
continental desert 84
continental drift 16, *20-21*
continental plates 12, *15, 17*, 19, 22, 25, 26
continental rise 52, *53*
continental shelf 52, *53*
continental slope 52, *53*
continents 16, 18-19, *19, 21*
convection currents 60
coral reef *52*
core 8, 14, *15*
corundum 37
creepers 77
Cretaceous period 28
crevasse *42*
crust 8, 14, *15*, 18, 28
crystals 28, 37, *37*

D

dam 56
Death Valley 76, 84, *84*
deciduous forest 80, *80*
deer *80*
deltas 41, *41*
deposition 30, 40, *41*
desert 84
destructive plate margins *17*, 19, 24
diamond *37*
dike *45*
dogwood bush 80
dolerite 33
dynamic meteorology 72

E

Earth 8-11, 12, *13,* 16
    core *see core*
    crust *see crust*
    density of 8
    formation of 10
    magnetic field of 20
    mantle *see mantle*
    mass of 8
    structure of 14, 15
    volume of 8
earthquakes 26, *27*
    epicenter of 26
    focus of 26
    recording of *26*
East Indies 50
El Chichón *74*
elephant *79*
emergents 77
englacial moraine *42*
epicenter 26
epiphytes 77
Equator 77, 82
erosion 39, 40, 45
Europe 80
evaporation *55*
Everest, Mount 28
exosphere 64, *64, 65*

F

fault 26, *26*
feldspars 32, 39
firn *42*
flint *37*
fold mountains 22, 23, 24
front *see weather*

92

94

Further reading

Baines, John D. *Protecting the Oceans*. Chatham, NJ: Raintree Steck-Vaughn, 1990.

Bromwell, Martyn. *Weather*. Danbury, CT: Franklin Watts, 1994.

Burchard, Elizabeth. *Earth Science – Geology: In a Flash*. Closter, NJ: Flash Blasters, 1994. (Exambusters Series)

Burroughs, William J. *Weather*. Alexandria, VA: Time-Life Books, 1996.

Catherall, Ed. *Exploring Soil and Rocks*. Chatham, NJ: Raintree Steck-Vaughn, 1990.

Catherall, Ed. *Exploring Weather*. Chatham, NJ: Raintree Steck-Vaughn, 1990.

Cosgrove, Brian. *Weather*. New York: Alfred A. Knopf Books for Young Readers, 1991.

Cuff, Kevin, et al. *Stories in Stone*. Berkeley, CA: Lawrence Science, 1995.

Dickey, John S., Jr. *On the Rocks: Earth Science for Everyone*. New York: Wiley, 1996.

Fuller, Sue. *Rocks and Minerals*. New York: Dorling Kindersley, 1995.

Ganeri, Anita. *The Oceans Atlas*. New York: Dorling Kindersley, 1994.

Grady, Sean M. *Plate Tectonics: Earth's Shifting Crust*. San Diego: Lucent Books, 1991.

Hall, Cally, and Scarlett O'Hara. *Earth Facts*. New York: Dorling Kindersley, 1995.

Kerrod, Robin. *The Weather*. Tarrytown, NY: Marshall Cavendish, 1994.

Mason, John. *Weather and Climate*. Morristown, NJ: Silver Burdett Press, 1991.

Mattson, Robert A. *The Living Ocean*. Springfield, NJ: Enslow Publications, 1991.

McIlveen, J. F. *Fundamentals of Weather and Climate*. New York: Chapman & Hall, 1991.

McMillan, Bruce. *The Weather Sky*. New York: Farrar, Straus & Giroux, 1991.

Neal, Philip. *The Oceans*. North Pomfret, VT: Trafalgar Square, 1993.

Pifer, Joanne. *EarthWise: Earth's Oceans*. Tucson, AZ: WP Press, 1992.

Potter, Tony. *Weather*. Jersey City, NJ: Parkwest Publications, 1992.

Sayre, April P. *Tundra*. New York: Twenty-First Century Books, 1994.

Tesar, Jenny. *Patterns in Nature: An Overview of the Living World*. Woodbridge, CT: Blackbirch Press, 1994.

Tesar, Jenny. *Threatened Oceans*. New York: Facts on File, 1992.

Wallace, Sally M. *Earthquakes*. Minneapolis: Carolrhoda, 1996.

Weather. New York: Dorling Kindersley, 1995.